ENVIRONMENTAL SCIENCE, ENGINEERING AND TECHNOLOGY

NATIONAL RADIATION INCIDENTS

LABORATORY GUIDE

ENVIRONMENTAL SCIENCE, ENGINEERING AND TECHNOLOGY

Additional books in this series can be found on Nova's website under the Series tab.

Additional E-books in this series can be found on Nova's website under the E-book tab.

ENVIRONMENTAL SCIENCE, ENGINEERING AND TECHNOLOGY

NATIONAL RADIATION INCIDENTS

LABORATORY GUIDE

MARTIN C. SHECKLEY
EDITOR

Nova Science Publishers, Inc.
New York

Copyright © 2011 by Nova Science Publishers, Inc.

All rights reserved. No part of this book may be reproduced, stored in a retrieval system or transmitted in any form or by any means: electronic, electrostatic, magnetic, tape, mechanical photocopying, recording or otherwise without the written permission of the Publisher.

For permission to use material from this book please contact us:
Telephone 631-231-7269; Fax 631-231-8175
Web Site: http://www.novapublishers.com

NOTICE TO THE READER

The Publisher has taken reasonable care in the preparation of this book, but makes no expressed or implied warranty of any kind and assumes no responsibility for any errors or omissions. No liability is assumed for incidental or consequential damages in connection with or arising out of information contained in this book. The Publisher shall not be liable for any special, consequential, or exemplary damages resulting, in whole or in part, from the readers' use of, or reliance upon, this material. Any parts of this book based on government reports are so indicated and copyright is claimed for those parts to the extent applicable to compilations of such works.

Independent verification should be sought for any data, advice or recommendations contained in this book. In addition, no responsibility is assumed by the publisher for any injury and/or damage to persons or property arising from any methods, products, instructions, ideas or otherwise contained in this publication.

This publication is designed to provide accurate and authoritative information with regard to the subject matter covered herein. It is sold with the clear understanding that the Publisher is not engaged in rendering legal or any other professional services. If legal or any other expert assistance is required, the services of a competent person should be sought. FROM A DECLARATION OF PARTICIPANTS JOINTLY ADOPTED BY A COMMITTEE OF THE AMERICAN BAR ASSOCIATION AND A COMMITTEE OF PUBLISHERS.

Additional color graphics may be available in the e-book version of this book.

Library of Congress Cataloging-in-Publication Data

National radiation incidents : laboratory guide / editors, Martin C. Sheckley.
    p. cm.
  Includes bibliographical references and index.
  ISBN 978-1-61324-666-5 (hardcover)
  1. Radioactive substances--Detection--Laboratory manuals. 2. Radiation--Measurement--Laboratory manuals. 3. Nuclear terrorism--United States--Safety measures--Laboratory manuals. 4. Emergency management--United States--Laboratory manuals. I. Sheckley, Martin C.
  TK9149.N38 2011
  363.325'565--dc23
                                                            2011015629

Published by Nova Science Publishers, Inc. † New York

Contents

Preface vii

Chapter 1 Method Validation Guide for Qualifying Methods Used by Radiological Laboratories Participating in Incident Response Activities 1
Environmental Protection Agency

Chapter 2 Radiological Laboratory Sample Screening Analysis Guide for Incidents of National Significance 79
Environmental Protection Agency

Index 125

PREFACE

The United States Environmental Protection Agency (EPA) is responsible for assessing the extent of environmental contamination and human health consequences in the event of a radiological incident such as a terrorist incident involving radioactive materials. Although the EPA will be mainly involved in the intermediate and recovery phases of an incident response, there also may be involvement in some activities in the early phase. This book describes project method validation guidance that a radioanalytical laboratory should comply with in order to validate methods used to process samples submitted during a radiological or nuclear incident, such as that caused by a terrorist attack. These responsibilities include response and recovery actions to detect and identify radioactive substances, and to coordinate federal radiological monitoring and assessment activities

Chapter 1- For a terrorist event such as a radiological dispersion device, the radionuclide(s) and the types and number of sample matrices that may be collected and analyzed can vary dramatically depending on the type of device used and radioactive material incorporated. The radioanalytical laboratories used to process the samples must not only be capable of identifying and quantifying the radionuclide(s) in various matrices, but they must also have the capacity to process a large number of samples in a short time (thousands of samples per week). Sufficient laboratory capacity is a balance of adequate facility processing areas and nuclear instrumentation, validated radioanalytical methods available, and trained staff.

Chapter 2- A response to a release of radioactivity to the environment likely will Most laboratories do not routinely screen samples under conditions found during an emergency response situation, such as from a radiological or nuclear incident of national significance (INS). Many of these samples are higher in activity and need to be accurately surveyed and prioritized for analysis based on direction from the Incident Commander (IC). This document describes methods that may be applied by personnel at a radioanalytical laboratory for screening of samples for radioactivity. The specific techniques described in this guide may be used to assess the gross α, β, or γ activity in samples that may have been contaminated as the result of a radiological or nuclear event, such as a radiological dispersion device (RDD), improvised nuclear device (IND), or an intentional release of radioactive materials into the atmosphere or a body of water or aquifer, or to terrestrial areas via mechanical or other methods. In the event of a major incident that releases radioactive materials to the environment, EPA will turn to selected radio- analytical laboratories to support its response and recovery activities. In order to expedite sample analyses and data feedback, the laboratories will need guidance occur in three phases that are generally defined in this document as: "early" (onset of the event to about day 4), "intermediate" (about day 4 to about day 30), and "recovery" (beyond about day 30). Each phase of an incident response will require different and distinct radioanalytical resources to address the different consequences, management, priorities, and requirements of a phase. Some of the more important radioanalytical laboratory issues germane to an incident response consist of radionuclide identification and quantification capability, sample load capacity, sample processing turnaround time, quality of analytical data, and data transfer capability. This guide emphasizes the laboratory screening of samples from the end of the early phase, through the intermediate phase, and into the recovery phase (but does not address the screening by initial responders).

In: National Radiation Incidents
Editor: Martin C. Sheckley

ISBN: 978-1-61324-666-5
© 2011 Nova Science Publishers, Inc.

Chapter 1

METHOD VALIDATION GUIDE FOR QUALIFYING METHODS USED BY RADIOLOGICAL LABORATORIES PARTICIPATING IN INCIDENT RESPONSE ACTIVITIES[*]

Environmental Protection Agency

ACKNOWLEDGMENTS

This manual was developed by the National Air and Radiation Environmental Laboratory (NAREL) of EPA's Office of Radiation and Indoor Air (ORIA).

Dr. John Griggs served as project lead for this document. Several individuals provided valuable support and input to this document throughout its development. Special acknowledgment and appreciation are extended to Dr. Keith McCroan, ORIA/NAREL; Mr. Daniel Mackney for his support of instrumental analysis, ORIA/NAREL; Dr. Lowell Ralston and Mr. Edward Tupin, CHP, both of ORIA/Radiation Protection Division; Ms. Schatzi Fitz-James, Office of Emergency Management, Homeland Security Laboratory

---

[*] This is an edited, reformatted and augmented version of United States Environmental Protection Agency Report EPA 402-R-09-006, dated June 2009.

Response Center; and Mr. David Garman, ORIA/NAREL. We also wish to acknowledge the peer reviews conducted by Carolyn Wong, David Burns, and Jack Bennett, whose thoughtful comments contributed greatly to the understanding and quality of the report. Numerous other individuals both inside and outside of EPA provided peer review of this document, and their suggestions contributed greatly to the quality and consistency of the final document. Technical support was provided by Dr. N. Jay Bassin, Dr. Carl V. Gogolak, Dr. Robert Litman, Dr. David McCurdy, Mr. Robert Shannon, and Dr. Anna Berne of Environmental Management Support, Inc.

ACRONYMS, ABBREVIATIONS, UNITS, AND SYMBOLS

(Excluding chemical symbols and formulas)

α	alpha particle
α	probability of a Type I decision error
AAL	analytical action level
AL	action level
APS	analytical protocol specification
β	beta particle
β	probability of a Type II decision error
Bq	becquerel (1 dps)
CL_{NC}	critical net concentration level
CSU	combined standard uncertainty
d	day
DAC	derived air concentration
DP	decay product(s)
dpm	disintegration per minute
dps	disintegration per second
DQO	data quality objective
DRC	derived radionuclide concentration
DWC	derived water concentration
EPA	[United States] Environmental Protection Agency
γ	gamma ray
g	gram
Gy	gray [unit of absorbed radiation dose in materials; 1 gray = 100 rad]
h	hour
IC	Incident Commander [or designee]

ISO	International Organization for Standardization
keV	thousand electron volts
L	liter
m	meter
MARLAP	*Multi-Agency Radiological Laboratory Analytical Protocols Manual*
MARS SIM	*Multi-Agency Radiation Survey and Site Investigation Manual*
MDB	minimum detectable bias
MDC	minimum detectable concentration
MeV	million electron volts
min	minute
mL	milliliter (10^{-3} L)
MQO	measurement quality objective
mrem	millirem (10^{-3} rem)
MSE	mean squared error
MV	method validation
MVRM	method validation reference material
PAG	protective action guide
pCi	picocurie (10^{-12} Ci)
PE	performance evaluation
PT	proficiency test/testing
QC	quality control
rad	unit of absorbed radiation dose in materials; 100 rad = gray
RDD	radiological dispersal device (i.e., "dirty bomb")
rem	roentgen equivalent man (traditional units; 1 rem = 0.01 Sv)
RSD	relative standard deviation
s	second
s_{Blanks}	standard deviation of blank sample net results
SI	International System of Units
Sv	sievert (1 sievert = 100 rem)
u_{MR}	required method uncertainty
φ_{MR}	relative required method uncertainty
y	year

RADIOMETRIC AND GENERAL UNIT CONVERSIONS

To Convert	To	Multiply by	To Convert	To	Multiply by
years (y)	seconds (s) minutes (min) hours (h) days (d)	3.16×10^7 5.26×10^5 8.77×10^3 3.65×10^2	s min h d	y	3.17×10^{-8} 1.90×10^{-6} 1.14×10^{-4} 2.74×10^{-3}
disintegrations per second (dps)	becquerels (Bq)	1	Bq	dps	1
Bq Bq/kg Bq/m^3 Bq/m^3	picocuries(pCi) pCi/g pCi/L Bq/L	27.0 2.70×10^{-2} 2.70×10^{-2} 10^{-3}	pCi pCi/g pCi/L Bq/L	Bq Bq/kg Bq/m^3 Bq/m^3	3.70×10^{-2} 37.0 37.0 10^3
microcuries per milliliter (µCi/mL)	pCi/L	10^9	pCi/L	µCi/mL	10^{-9}
disintegrations per minute (dpm)	µCi pCi	4.50×10^{-7} 4.50×10^{-1}	pCi	dpm	2.22
cubic feet (ft^3)	cubic meters (m^3)	2.83×10^{-2}	cubic meters (m^3)	cubic feet (ft^3)	35.3
gallons (gal)	liters (L)	3.78	L	gal	0.264
gray (Gy)	rad	10^2	rad	Gy	10^{-2}
roentgen equivalent man (rem)	sievert (Sv)	10^{-2}	Sv	rem	10^2

Note: Traditional units are used throughout this document instead of International System of Units (SI) units. Protective Action Guides (PAGs) and their derived concentrations appear in official documents in the traditional units and are in common usage. Conversion to SI units will be aided by the unit conversions in this table. Conversions are exact to three significant figures, consistent with their intended application.

1.0. INTRODUCTION

The United States Environmental Protection Agency (EPA) is responsible for assessing the extent of environmental contamination and human health consequences in the event of a radiological incident such as a terrorist incident involving radioactive materials. Although EPA will be mainly involved in the intermediate and recovery phases of an incident response, there also may be involvement in some activities in the early phase. For a terrorist event such as a radiological dispersion device, the radionuclide(s) and the types and number of sample matrices that may be collected and analyzed can vary dramatically depending on the type of device used and radioactive material incorporated. The radioanalytical laboratories used to process the samples must not only be capable of identifying and quantifying the radionuclide(s) in various matrices, but they must also have the capacity to process a large number of samples in a short time (thousands of samples per week). Sufficient laboratory capacity is a balance of adequate facility processing areas and nuclear instrumentation, validated radioanalytical methods available, and trained staff.

In order to make proper assessments and decisions in the event of a radiological incident, EPA will utilize only qualified radioanalytical laboratories that have the capability, capacity and quality needed to process samples taken from affected areas. Analytical protocol specifications (APSs), including measurement quality objectives (MQOs), will be preestablished to define the expected quality of the data for incident situations. The objective of this document is to establish systematic and objective methodologies and acceptance criteria for validating analytical methods, based on the stated quality requirements of a specific incident-response project, such as recovery from a radiological dispersal device. Laboratories developing new methods and operational protocols should review the detailed guidance on recommended radioanalytical practices found in current editions of MARLAP and MARSSIM.

Several radiological sample analysis guides for incident response have been developed that provide information on the expected radionuclides of concern and MQOs to make decisions relative to sample processing priorities for the water, air particulate filter, and soil/solid matrices. As part of the laboratory qualifying process, laboratories must demonstrate their ability to meet the APSs and MQOs for the methods used to analyze each radionuclide and sample-matrix combination. EPA will require an initial project method validation and a subsequent participation in a performance evaluation (PE) program as a means to demonstrate that the methods used by a laboratory are capable of meeting the MQOs for incident response applications. For incident-response applications, project method validation will be required and applied to methods currently being used by the laboratories, including EPA Safe Drinking Water Act required methods, as well as to newly developed methods and methods that have been modified for incident response. Project method validation and participation in a PE program will be required for gross alpha and beta screening methods as well.

In this document, the term "project method validation" is synonymous with "incident response method validation."

2.0. METHOD VALIDATION DESCRIPTION

The *Multi-Agency Radiological Laboratory Analytical Protocols Manual* (MARLAP) Chapter 6 discusses two distinct applications of method validation: general method validation and project method validation. General method validation is the process of demonstrating that a method is suitable for its general intended use, such as routine radioanalytical processing of samples for the determination of environmental levels of a given radionuclide. For general method validation, the methods would address internal measurement quality objectives, and typical sample matrix constituents and nominally interfering concentrations of expected chemical and radionuclide interferences. EPA has developed a draft general method validation process document (EPA 2006, *Validation and Peer Review of U.S. EPA Radiochemical Methods of Analysis*) that covers the method validation parameters for radioanalytical methods. That document provides guidance to satisfy EPA requirements for general method validation for measurement uncertainty, method bias and trueness, precision, detection capability, analyte concentration range, specificity and ruggedness.

In contrast, this document provides guidance on project method validation applicable to methods for processing samples during a response to a radiological incident, including radiological incidents of national significance. Project method validation demonstrates that a method is capable of meeting project-specific MQOs (in other words, a required method uncertainty at a specific radionuclide concentration). The method selected for a project needs to address specific sample matrix characteristics, chemical and radionuclide interferences, special sample preparation requirements, sample-processing turnaround times, and MQOs defined in an analytical protocol specification (APS). This document addresses the method validation expectations for an incident response for the MQOs of the required method uncertainty and the required minimum detectable concentration (MDC). The method validation procedures for the method uncertainty MQO follow the guidance provided in MARLAP Chapter 6. As discussed in MARLAP, the principal MQO is the required method uncertainty at an action level. Although the MDC MQO normally would not be specified as an MQO for incidence response applications, this document provides method validation guidance for a "required MDC" MQO.

Even though a laboratory has a method that has undergone general method validation, use of the method for the incident response application will require project method validation. The degree of effort and required level of project method validation will depend on the degree of method development or use, and the MQOs of the project, as included in the APSs.

Proper planning is critical for successful method validation because many method validation parameters must be considered, evaluated and documented. Method development and method validation generally are not separate processes. The types of experiments conducted during method development and the types of tests performed during method validation have many similarities.

3.0. METHOD DESCRIPTION

The components of a method or measurement process requiring validation should be clearly described. Generally, a laboratory method includes all physical, chemical and radiometric processes conducted at a laboratory in order to provide an analytical result. The processes for radiochemical methods may include sample preparation or dissolution, chemical separations, preparation of sample test sources, nuclear counting, analytical calculations, data review and qualification, and data reporting (MARLAP Chapter 6). Method validation efforts should evaluate all process components combined. Some radiochemical methods may also include procedures for sampling (e.g., methods for radon in air analysis or for volatile radioactive organic compounds in soils and other solid matrices), in which case the sampling procedures should be included in the validation tests. The measurement process components validated, and the combination of procedures comprising a method, must be clearly and completely stated.

The purpose of a method (i.e., measurement objectives) and the intended use of the data must be clearly defined. In addition, method scope and applicability must be well defined and clearly described and consistent with the documented performance of the method. These measures will help minimize misapplication by the users. Method scope and applicability include the following:

- The measurement process components validated (e.g., sample preparation, dissolution, chemical isolation, precipitation, final product for counting, radiation measurement process, etc.)
- The nature (chemical-physical form, type of radiation and quantity measured) of the radionuclides and matrices (chemical and physical form) studied
- The range of analyte concentration levels for which the method is claimed to be suitable
- A description of any known limitations and any assumptions upon which a method is based (e.g., radiological and non-radiological interferences, minimum sample size, etc.)
- A description of how the method and analytical parameters chosen meet the measurement quality objectives for the intended application, when applicable
- Aliquant sample size for processing

4.0. METHOD PERFORMANCE CHARACTERISTICS

The performance characteristics of a radiochemical method that may be evaluated in method validation include:

- Method uncertainty at a specific radionuclide concentration (action level)
- Detection capability (minimum detectable concentration)
- Bias/trueness
- Analyte concentration range
- Method specificity
- Method ruggedness

A brief discussion of each of these performance characteristics will be covered in the following sections. For more detailed information on a characteristic, the reader is referred to MARLAP (Chapters 3 and 6); EPA (2006) *Validation and Peer Review of U.S. EPA Radiochemical Methods of Analysis*; EURACHEM Guide (1998) *The Fitness for Purpose of Analytical Methods, A Laboratory Guide to Method Validation and Related Topics*; ISO 17025; and ANSI N42.23.

4.1. Method Uncertainty

MARLAP defines method uncertainty as follows:

Method uncertainty refers to the predicted uncertainty of the result that would be measured if the method were applied to a hypothetical laboratory sample with a specified analyte concentration. Although individual measurement uncertainties will vary from one measured result to another, the required method uncertainty is a target value for the individual measurement uncertainties, and is an estimate of uncertainty (of measurement) before the sample is actually measured.

Method uncertainty can be thought of as an estimate of the expected analytical standard deviation at a specified radionuclide concentration. For certain projects, including incident response, a required method uncertainty should be specified. An example of a required method uncertainty specification would be "...at a ^{137}Cs soil concentration of 10 pCi/g, the required method uncertainty is 1 pCi/g." In many applications, including incident response laboratory analyses, the specified radionuclide concentration is referred to as the analytical action level (AAL) and may be based on either incident- specific, risk-based or regulatory mandated value, such as a protective action guide (PAG) as presented in the *Radiological Laboratory Sample Analysis Guides for Incidents of National Significance*. Radioanalytical results from an incident response are compared to action level concentrations, and thus it is very important to have results that are of sufficient quality to support decisions to be made. Specifying a required method uncertainty at the AAL ensures the data quality needed to make decisions.

To be consistent with MARLAP, certain nomenclature for the required method uncertainty is used for incident response applications. The notation "u_{MR}" is specified for the absolute required method uncertainty at or below the action level and has units of activity or activity concentration that match the AAL value. Above the action level, a relative required method uncertainty φ_{MR}, defined as the u_{MR}/AAL, is specified (φ_{MR} is unitless). For the ^{137}Cs soil example provided above, nMR would be equal to:

$$= \frac{1 \text{ pCi/g}}{10 \text{ pCi/g}} = 0.10 \ (10\%) \text{ at concentrations greater than 10 pCi/g}$$

Method uncertainty should not be confused with measurement uncertainty, which MARLAP and the International Organization for Standardization (ISO) (1993 a) defined as:

"Parameter associated with the result of a measurement that characterizes the dispersion of the values that could reasonably be attributed to the measurand."

Each radioanalytical method that will be used by a laboratory processing samples from an incident response will be evaluated to determine if its uncertainty meets the required method uncertainty. The result of each test sample processed during the method validation process is compared to the limits of acceptability established for the specific validation level, i.e., a multiple of the required method uncertainty. A method will be considered acceptable if it meets the method validation criteria provided in Section 5.4 for the appropriate level of validation. Derived radionuclide concentrations (DRCs) corresponding to the AAL for the water, air filter and soil matrices can be found in the *Radiological Laboratory Sample Analysis Guide for Incidents of National Significance* (see Appendix A for summary tables). For project method validation, MARLAP recommends that the uncertainty of a method be evaluated at or near an action level radionuclide concentration. In the absence of defined AALs and required method uncertainties (either by the Incident Commander [IC][1] or other project manager), default AALs and corresponding required method uncertainties for the method validation test samples provided in Sections 5.2 and 5.4 can be used. The default values in Appendix A may be considered "acceptable" starting levels. The IC may develop and require other AALs and required method uncertainties. If so, the IC should verify whether the laboratory can meet the new method uncertainty requirements for the updated AALs.

4.2. Detection Capability

In some cases, the detection capability of a method, rather than the required method uncertainty, is the important MQO of a project. Detection capability for this guide uses the concept of the minimum detectable concentration (MDC) or minimum detectable value. MARLAP defines the minimum detectable value of the analyte concentration in a sample as:

An estimate of the smallest true value of the measurand that ensures a specified high probability, $1 - \beta$, of detection.[2]

For radioanalytical processes, the probability of detection $(1 - \beta)$ of 0.95 is commonly used. The definition of the minimum detectable value presupposes that an appropriate detection criterion has been specified, i.e., "critical net concentration" for this document. This approach assumes that the measured radionuclide net concentration in a sample will be above the critical net concentration 95% of the time if the true concentration is equal to the MDC.

MARLAP (Chapter 20) provides a detailed discussion on how to calculate the critical net concentration and MDC using a number of equations for various applications. The equations provided in MARLAP calculate estimates of these method detection parameters for a given method based on either a measured signal response of a single blank sample or from a population of sample blanks that have been processed by the method under evaluation. For those applications when a required MDC for a method has been specified as an MQO, the detection capability of the method should be evaluated during method validation.

4.3. Bias and Trueness

Bias refers to the overall magnitude of systematic errors associated with the use of an analytical method. The presence of systematic errors can be determined only by comparison of the average of many results with a reliable, accepted reference value. Method bias may be estimated by measuring materials whose composition is reasonably well known, such as reference materials, by comparing results to those from at least one alternate method or procedure, or by analyzing spiked materials.

ISO (1993a) defines bias as:

"[T]he mean value that would result from an infinite number of measurements of the same measurand carried out under repeatability conditions minus a true value of the measurand."

According to MARLAP (Chapter 6), bias typically cannot be accurately determined from a single result or a few results because of the uncertainty in the measurement process to determine the measurand. Bias is normally expressed as the absolute or relative deviation of the average of a group of samples from the "true" or "known" value. Since it is a calculated estimate, a bias should be reported with a combined standard uncertainty and include the number of data points used to calculate the bias.

It is assumed that the mean response of the method is essentially a linear function of analyte concentration over the useful range of the method. As defined in MARLAP, "this function can be characterized by its *y*-intercept, which reflects the mean response at zero concentration, and its slope, which reflects the ratio of the change in the mean response to a change in sample analyte concentration." The "absolute bias" of a method can be thought of as the difference between the average concentration of the radionuclide at the y-intercept and the true concentration of zero.

The IC will specify a method bias limit as an APS when method bias is considered an important method performance characteristic for the method or a quality parameter for the project. Method bias must be evaluated during method development, general and project method validation processes, and subsequently, the processing of batch quality control (QC) samples processed with the incident response samples (MARLAP Chapter 7).

The method uncertainty acceptance criteria provided in MARLAP (Chapter 6), as well as for project method validation in Section 5.4, assume that laboratories would not use a method that has a significant bias. When a method has excessive bias, the method validation test results for the required method uncertainty will be unacceptable. Appendices D and E provide information on bias evaluation methods as related to the method validation acceptance criteria.

4.4. Analyte Concentration Range

The analyte concentration range of a method is a method performance characteristic that defines the span of radionuclide activity levels, as contained in a sample matrix, for which method performance has been tested and data quality deemed acceptable for their intended use. However, not all sample matrices encountered during an incident response will have preestablished analytical action levels with corresponding required method uncertainty values or required MDCs. Therefore, incident response method validation must be sufficiently flexible to address not only those typical sample matrices (liquids, air filters, swipes, and soil/solids) for which there are action levels, but also those matrices for which there are no specified action levels. The subsequent subsections discuss the analytical concentration range options for method validation for both situations. For both options, the method is to be tested at a low, mid and upper validation test concentration/activity except when noted.

4.4.1. Derived Radionuclide Concentrations Corresponding to Established Action Levels

MARLAP (Chapter 6) recommends that a method be validated at the expected action level for a radionuclide and matrix combination. Therefore, an analyte concentration range should include either an established regulatory limit or a defined action level, typically near the midpoint of the radionuclide activity (concentration) range for a project. For a radiological incident response application, the established AAL would normally be a derived radionuclide concentration corresponding to a PAG or a risk-based dose as designated by an agency representative. There may be four or five action levels for the various matrices contaminated, with the range of concentrations as great as four orders of magnitude. Derived radionuclide concentrations for the various established AALs have been generated for water and air-filter matrices (AALs for soils/sediments and building materials are being developed), and can be found in the *Radiological Laboratory Sample Analysis Guide for Incidents of National Significance* series. Summary tables of the established AALs for these three matrices can be found in Appendix A. AALs vary according to the matrix, phase of the incident and applied PAG.

At most laboratories, samples that have been screened and found to contain very high radionuclide activities probably will be subdivided (when possible) prior to specific radioanalytical processing. Also, samples requiring multiple radionuclide analyses should be subdivided. Both situations will result in lower-activity subsamples (aliquants) for processing. For example, when a sample that has a very high radionuclide concentration or activity is received by a laboratory, the sample likely will be subdivided (possibly into five parts) so that a different radioanalytical pathway for each radionuclide may be performed in parallel. For aqueous and soil samples, the radionuclide concentration of the radionuclides in the aliquants would be the same as the original aqueous and soil concentrations, but the aliquant activity available for processing will be reduced proportionally from the original sample size. For air-filters, the total activity on the filter matrix represents the activity in a volume of air collected. For swipe samples, the activity on the sample represents the activity removed from a surface area swiped. Air filter and swipe samples may be digested prior to radiochemical processing and the digestate volume generated represents the total activity on the original sample for the air volume collected or surface area swiped. Aliquanting these digestates to obtain a lower subsample activity or for multiple analyses is also a common practice. Thus, it is important to know the exact fraction of the original sample taken so that the analysts know that a sufficient sub-sample quantity has been processed to ensure that the MQOs have been met.

When developing incident response methods for high activity samples, it is important to note that the analytical concentration range and detection capability specifications will be significantly higher than what is usually found in normal procedures for environmental monitoring sample processing. This difference in concentration range should be emphasized in the procedure's scope.

4.4.2. Default Analytical Action Levels

Established AALs based on PAG-derived radionuclide concentrations may not be available for all matrices encountered in an incident response, such as concrete or asphalt. In the absence of established PAG action levels, default AALs may be used for the validation test concentration or activity levels. Section 5.4 provides guidance on selecting default AALs applicable to method validation for the three general matrix categories of liquids, air sampling media/swipes and solids. The default AALs approximate the expected derived radionuclide activity level for a sample volume or mass for a 100-mrem or 10^{-4} risk-based AAL. These AAL levels were chosen because they can be conveniently scaled to other possible project-specific AALs for the various matrices. For example, if a specific project had an AAL at 20 mrem (one-fifth of a 100 mrem AAL), the table values for the AALs can be scaled down simply by dividing the listed values by five.

4.5. Method Specificity

MARLAP defines "method specificity" as "the ability of the method to measure the analyte of concern in the presence of interferences." EURACHEM (1998) defines selectivity or specificity as "the ability of a method to determine accurately and specifically the analyte of interest in the presence of other components in a sample matrix under the stated conditions of the test."

By extension, this guide defines method specificity as:

The ability to correctly identify and quantify the radionuclide(s) of interest in the presence of other interferences in a sample under stated conditions of the test.

Method specificity should be evaluated during method development and the general and project method validation processes for the applicable matrices, radionuclide(s) of interest and known interfering radionuclides. Method specificity may be evaluated during method validation by analyzing:

- Matrix samples that have been characterized in terms of radionuclide and chemical constituent content;
- Appropriate matrix blanks; and
- Matrix blanks spiked with interferences.

Each specific sample matrix should be tested for method specificity, e.g., concrete, asphalt, soil, etc. Matrix samples and blanks should be chosen to be as representative of the target matrix as is practical. When possible, matrix blanks should contain the chemical species and potential interfering radionuclides, other than the radionuclide(s) of interest, at concentrations that are reasonably expected to be present in an actual sample. Each of the three options to determine method specificity may provide insight into the relative degree of expected quantitative effect that the interferences will have on the identification and quantification of the radionuclide(s) of interest at different concentrations.

Method specificity is typically expressed qualitatively and quantitatively. A radiochemical method specificity statement would include descriptions of parameters, such as:

- Expected radionuclide and chemical interferences
- Effects of the interfering substances on the measurement process
- Measurement information that substantiates the identity of the analyte (e.g., half-life, or decay emission and energy)
- Effects of oxidation or molecular state of the target or interfering radionuclides
- Chemical processes that can remove interfering materials (e.g., ion exchange, solvent extraction)
- Summary of results from analysis of standards, reference materials and matrix blanks

4.6. Method Ruggedness

MARLAP defines "method ruggedness" as "the relative stability of method performance for small variations in method parameter values." EURACHEM (1998) discusses the concept of method ruggedness and robustness interchangeably. Ruggedness is a measure of how well a method's performance stands up to less than perfect implementation. In any method there are certain steps which if not carried out sufficiently, exactly or carefully may have a significant effect on method performance and the reliability of the results. Typically, these critical steps are identified during the method development process, and annotations are made in the method description that provide limiting conditions and an allowable range of application. It is advantageous to identify the variables in the method that have the most significant effect on the analytical results so that they are closely controlled. Ruggedness or robustness tests have been developed which involve experimental designs for examining method performance when minor changes are made in operating steps or in some cases environmental conditions (EPA 2006, *Validation and Peer Review of the U.S. EPA Radiochemical Methods of Analysis*). The tests involve making deliberate variations to the method, and investigating the subsequent effect on performance.

An example of method ruggedness is the adjustment of pH during the separation of strontium from calcium in the analysis of milk for ^{90}Sr. The pH of the milk is buffered at 5.4 and disodium EDTA is added prior to passing the solution through a cation exchange column. The calcium will effectively complex with EDTA at this pH, forming an anion, while the strontium remains a cation. A pH lower than about 5.2 will not provide enough EDTA anion to effectively complex calcium, and a pH greater than about 5.5 will begin to effectively complex strontium. Thus for this analysis method, ruggedness deals with pH control in the range of 5.2 to 5.5.

Method ruggedness is typically evaluated during method development and prior to method validation. Therefore, no specific tests for ruggedness will be included in this document for project method validation of the radioanalytical methods used for incident response.

5.0. INCIDENT RESPONSE METHOD VALIDATION GUIDANCE, TESTS, AND REQUIREMENTS

This section provides guidance, specific tests and minimum requirements for project method validation for methods used to process samples from a radiological incident. This section addresses the following selected method performance characteristics:

- Method specificity
- Analyte concentration range
- Method validation levels for testing the required method uncertainty
- Verification of required detection limit specification

Discussion of matrix considerations and method bias tests are also included in this section. Before initiating the method validation process, a validation plan should be prepared that incorporates the various guidance and requirements specified in this section and Section 6, Method Validation Documentation.

5.1. Method Specificity

Method specificity is evaluated during general method validation for normal routine applications, e.g., environmental surveillance programs. During general method validation, the method should be evaluated for the applicable matrices and radionuclide(s) of interest and known interfering chemical constituents and radionuclides over a typical expected range. For some incident response applications, method validation testing for method specificity may be more focused than general method validation. Incident response scenarios may involve one or many radionuclides and a multitude of matrices. To ensure method specificity for the incident response application have been met, the proficiency testing (PT) samples used for incident response method validation should contain the known or expected concentration levels of the matrix chemical species and potential interfering radionuclides. Adequate method specificity during project method validation should be evaluated by analyzing:

- Matrix PT samples that have been characterized in terms of expected radionuclide and chemical constituent content;
- Appropriate matrix blanks containing the applicable radionuclide and chemical interferences; and
- Matrix blanks spiked with interferences.

During method development, decontamination factors[3] should be evaluated for the more commonly expected radionuclide interferences so that the final method can improve method performance and adequately address radionuclide interferences. Also, the concentration of the interfering radionuclides should be added during method development at their 100 mrem AAL-derived radionuclide concentrations. Matrix blank results having no absolute bias would indicate adequate method specificity. Excessive absolute or relative bias, erroneous chemical or radiotracer yields, or possibly excessive method uncertainty may be indications of inadequate method specificity.

5.2. Analyte Concentration Range

The radionuclide concentration range applicable to method validation for radiological incident response should extend from a lower bound (~0.5 AAL) to an upper bound (3 AAL) that are both a multiple of an incident response action level. For radiological incident response applications, the analyte concentration range for the method validation process and validation test levels should be established based on the established PAG or risk-based derived radionuclide concentrations as designated by a representative of the responsible government agency. If the laboratory has not been provided with action levels by the IC, default values listed in Table 1 may be used. Also, derived radionuclide concentrations for the various AALs for water, air filter and soil matrices can be found in the *Radiological Laboratory Sample Analysis Guide for Incidents of National Significance*. Summary tables of the AAL-derived radionuclide concentrations for the water and air filter matrices can be found in Appendix A[4] of this document. These tables list the expected AALs for various media mainly for the late intermediate and recovery phases, but also for the early phase. The AALderived radionuclide concentrations will vary according to the matrix, phase of the incident and applied PAG. For air filters, an activity per sample corresponding to an AAL concentration for an assumed air volume sampled should be used.

Table 1. Default Analytical Action Levels for General Matrix Categories

Matrix Category	Size Assumptions for Values	Default Test Level Activity in Each Sample Aliquant (Total pCi) [1]		
		Alpha (^{241}Am)	Pure Beta (^{90}Sr)	Gamma (^{60}Co)
Liquids	5-mL Screen	2.0	12	33
	100-mL Nuclide-Specific	40	240	660
Air Sampling Media/Swipe	68 m^3 Screen	22	1,900	8,400
	68 m^3 (4 aliquants [2]) Nuclide-Specific	5.5	480	2,100
Solids – soil, etc.	2 g	TBD [3]		
	100 g			
	500 g			

[1] Test-level activity corresponds closely to 100-mrem dose-derived concentration values for water and 10^{-4} risk-based DAC for air (Appendix A). The table values were calculated for the noted radionuclides. To calculate air sampling default AALs for 10^{-6} risk-based applications, the 10^{-4} risk-based values in the table can be scaled down by a factor of 100. Table values for the solids and soil are pending.

[2] Test-level activity assumes that the air filter has been split into four aliquants after sample digestion.

[3] TBD: To be determined pending development of *Radiological Laboratory Sample Analysis Guide for Incidents of National Significance–Radionuclides in Soils and Solids*.

The validation test concentration/activity values should be adjusted to reflect the typical sample aliquant size that would be analyzed. In some cases, the original sample may be aliquanted directly, but in other cases the sample must be completely digested before sample aliquanting. When the radionuclide(s) identity is known, the number of aliquants may be small, but when the identity is not known, the number of aliquants may be three or more depending on the decay particle emission type.

In the absence of established PAG AALs, default AALs may be used. Default AALs (activity per sample aliquant) for three general matrix categories and radionuclide emission type are provided in Table 1. Default AALs can be used for similar matrix categories, as discussed in Section 5.3

(for all solutions, use water; for swipes, use air-filter materials; and for pulverized concrete, use soil). It should be noted that these default AALs and associated required method uncertainty values (Table 4) for the stated general matrix categories do not have a dosimetric basis but may be considered adequate for method validation purposes.

For each matrix category, default AALs are provided for both screening and specific radionuclide methods requiring validation. The test level values stated in Table 1 were calculated using the DRC for the 100-mrem AAL values for the gross alpha, beta, and gamma screening levels and for ^{241}Am, ^{90}Sr, and ^{60}Co for the specific alpha and beta/gamma radionuclide categories. The default AALs have been adjusted to reflect the typical sample aliquant size (column 2 of Table 1) that would be analyzed by a laboratory.

For practical reasons and to prevent potential laboratory/instrumentation contamination and radiological safety issues, the test levels for incident response method validation purposes are limited to three levels related to the designated (established) PAG AALs (derived radionuclide concentrations) or default AALs. The use of three test levels is consistent with the specifications indicated in Table 3 and Section 5.4. For incident response method validation, the validation test levels are denoted as lower, mid, and upper. For method validation levels B, C, D, and E (Section 5.4), the recommended three concentration test levels for the replicate PT samples are presented in Table 2. The lower level test level of ~0.5 AAL was chosen to avoid detectability issues that could occur at lower test concentrations. The mid test level corresponds to the established PAG AAL or default AAL test level. It is assumed that a laboratory will use the same sample aliquant size and counting time to analyze the test samples for all three test-level concentrations (values typically selected to meet the required method uncertainty at the mid test level, or AAL).

Table 2. Method Validation Test Concentrations

Test Level	Relative Concentration
Upper	~3 AAL
Mid	~1 AAL
Lower	~0.5 AAL

5.3. Matrix Considerations

For method validation, the method under consideration shall address a specific radionuclide and matrix combination. In many applications, a matrix may be described by a general name or type, such as water or air particulate. However, when developing and documenting the applicability of a method, a description of the sample matrix should be specific and address possible variations in the matrix that may be encountered when such will impact method performance. In addition, validation of a method applies only to the specifically defined matrix described in the method validation plan, which must be consistent with the matrix description in the method applicability statement. Listed below are some specific matrices that may be encountered for radioanalytical processing during an incident:

- Liquids
    - Fresh water
    - Surface water
    - Groundwater
    - Rain
    - Salt/brackish water
    - Aqueous suspensions
    - Aqueous solutions
    - Sewer and water treatment effluents or discharges – Collection of volatiles
    - Organic liquids
    - Liquids generated during decontamination activities
- Air sampling media
    - Glass fiber, cellulose, acetate filters
    - Charcoal canisters or loose particles
    - Molecular sieve
    - Silica gel
- Swipes
    - Glass fiber, cellulose, acetate filter paper
- Solids
    - Soil, sediment, stone, sod, vegetation, wood
    - Manufactured/ construction
    - Concrete, asphalt, brick, ceramics, plaster, plastics, metals, clothes, paper, stone, wood, etc.

- Sludges
- Sewer and water treatment
- Solids generated during decontamination activities

5.4. Method Validation Levels for Testing the Required Method Uncertainty

The primary method validation approach used in this document follows the concepts presented in MARLAP Chapter 6 for the required method uncertainty MQO. The MARLAP method validation approach and validation acceptance criteria assume that the laboratory method being validated has no significant bias. However, this may not always be the case. Appendix D provides an insight into the effect of method bias on the probability of failing the MARLAP validation acceptance criteria.

An alternate approach that may be used to determine if a method has acceptable method validation performance is presented in Appendix E. This approach is based on the mean squared error (MSE) or root mean squared error concept and has a greater power to detect excessive imprecision or bias in many cases.

If a method fails to meet the method validation acceptance criteria as presented in the subsequent sections, the laboratory should:

- Evaluate the possible reasons for the failure;
- Identify the root causes for the failure; and
- Update the method with the appropriate corrections or additions to ensure the method will meet the specified MQOs.

The updated method must go through another validation process using the same requirements applied to the first attempt at method validation.

5.4.1. Method Validation Requirements Based on MARLAP Concepts

Similar to the MARLAP (Chapter 6) graded approach to project method validation, there are four proposed tiers or "levels" of method validation (Levels B, C, D, E) to demonstrate a method's capability of meeting the required method uncertainty MQO applicable to a radiological incident. For this guide, the MARLAP method validation Level A for the *same* radionuclide and matrix combination has been combined with validation Level B (see Table 3). The level(s) of method validation needed should be designated by the IC.

The laboratory will select a method based on various operational aspects and the status of it's existing methods to meet the required method uncertainty u_{MR} or φ_{MR} specification for a designated (established) AAL (Appendix A) or a required method uncertainty for a default AAL (Section 5.2). The u_{MR} is specified in the units of the AAL. The φ_{MR} is a fractional unitless value (e.g., 0.13) and is calculated by dividing the u_{MR} by the AAL. Appendix A contains tables listing the required method uncertainties for screening and nuclide-specific methods for certain established AALs and sample matrices related to a potential radiological incident.

The four levels (B-E) of method validation for testing compliance with the required method uncertainty using specified PT samples cover the following:

Level B – Existing methods for radionuclide and matrix combinations for same, similar or slightly different matrices (internal PT samples);

Level C – Existing methods that require modification to accommodate matrix differences (internal or external PT samples);

Level D – Adapted or newly developed methods (internal and external PT samples); and

Level E – Adapted or newly developed methods using method validation reference materials (method validation reference materials).

During the method validation process, the laboratory shall evaluate the method as to the required method uncertainty and relative bias for the three test concentrations (Table 2) for the specified method validation level, as well as the absolute bias through the use of at least seven blanks (Section 5.6 and Appendix E). The acceptable performance of a method to meet the required method uncertainty will vary according to the level of validation as described in subsequent subsections. It should be noted that the probability of acceptable performance for meeting a required method uncertainty specification is dependent on the magnitude of existing method bias. The greater the magnitude of the method bias, the more likely the method will not meet the required method uncertainty specification. If excessive bias is measured during the method validation process, the method should be revised to eliminate the bias as much as possible.

Table 3. Method Validation Requirements and Applicable to Required Method Uncertainty

Validation Level [1]	Application	Sample Type	Acceptance Criterion [2]	Levels [4] (Concentration)	Replicates [3]	# of Analyses
B	Existing Method Radionuclide – Same, Similar or Slightly Different Matrix	Internal PT	Measured Value Within ±2.8 u_{MR} or ±2.8 φ_{MR} of Validation Value	3	3	9
C	Similar Matrix: New Application	Internal or External PT	Measured Value Within ±2.9 u_{MR} or ±2.9 φ_{MR} of Validation Value	3	5	15
D	Adapted, Newly Developed, Rapid Methods	Internal or External PT	Measured Value With in ±3.0 u_{MR} or ±3.0 φ_{MR} of Valida-tion Value	3	7	21
E	Adapted, Newly Developed, Rapid Methods	Method Validation Reference Materials	Measured Value Within ±3.0 u_{MR} or ±3.0 φ_{MR} of Validation Value	3	7	21

[1] MARLAP method validation Level A for the same radionuclide and matrix combination has been included into validation Level B.
[2] The acceptance criterion is applied to each analysis/test sample used for method validation, not the mean of the analyses. u_{MR} and φ_{MR} values are the required absolute and relative method uncertainty specifications stipulated in the *Radiological Laboratory Sample Analysis Guide for Incidents of National Significance* for gross screening concentrations and quantification of individual radionuclide concentrations in various matrices. The acceptance criteria are chosen to give a false rejection rate of ~5% when the

measurement process is unbiased, with a standard deviation equal to the required method uncertainty (u_{MR} or φ_{MR}). The stated multiplier (k = 2.8, 2.9, 3.0) for the required method uncertainty was calculated using the formula $k = z_{0.5} + 0.5(1-\alpha)^{1/N}$ where N is the number of measurements, α is the desired false rejection rate, and, for any p, z_p denotes the p-quantile ($0 < p < 1$) of the standard normal distribution (MARLAP Appendix G, Table G. 1). The u_{MR} or φ_{MR} values are provided in Appendix A or Table 4.

[3] For certain matrices, not all samples in a given test level can be spiked with the same known radionuclide activity or concentration. In such cases, the measured activity or concentration in the test sample should be compared to the known value for that test sample.

[4] At least seven blank samples should be analyzed as part of method validation but are not considered part of the three required concentration test levels.

For radiological incident response applications, the analyte concentration range for the method validation process, and thus the validation test levels, should be established based on the established PAG or risk-based derived radionuclide concentrations as designated by an agency representative. Derived radionuclide concentrations corresponding to the various established AALs (PAG or risk- based) for water, air particulate and soil matrices have been provided in the *Radiological Laboratory Sample Analysis Guide for Incidents of National Significance* and summarized in the Appendix A. In the absence of validation test activity levels based on established AALs (PAG or risk-based), default AALs specified in Section 5.2 may be used. Validation test activity/sample levels (designated established AALs or default AALs) are to be used in conjunction with all method validation levels stated in Table 3 and the required method uncertainty values for the radionuclide and matrix combinations provided in the *Radiological Laboratory Sample Analysis Guide for Incidents of National Significance* (Appendix A) or in Section 5.2.

A method is considered validated for a project when it has met the required method uncertainty acceptance criteria stated in Table 3 and the acceptance criteria for other method characteristics such as bias and required MDC, as may be stated by the IC. When the required method uncertainty specifications in Table 3 are met for default AALs, it will be assumed that the method has met the required method uncertainty acceptance criteria for all PAG or risk-based action levels above the default AALs.

All method validation levels require replicate samples at three different validation test concentration/ activity levels below, at, and above the derived radionuclide concentration corresponding to an AAL (designated, established, or default). To ensure testing for sufficient method specificity, the known concentration levels of potentially interfering radionuclides should be included in the test samples.

5.4.2. Required Method Uncertainty Acceptance Criteria

For all four method validation levels for method uncertainty, acceptable method validation is determined by comparing each test sample result for a given test concentration or activity with the required method uncertainty specification ($k \times u_{MR}$ or $k \times \varphi_{MR}$) provided in Table 3. The values for u_{MR} or φ_{MR} are stipulated in the *Radiological Laboratory Sample Analysis Guide for Incidents of National Significance* (Appendix A) for the three basic incident response matrices of water, air filter/swipes, and soil. A "k" value can be either 2.8, 2.9, or 3.0. It should be noted that the required method uncertainty specification and AAL-derived radionuclide concentrations may vary according to the matrix, the phase of the incident response and applied PAG.

Appendix B provides examples for testing a method's acceptability to meet validation Level D for an established PAG AAL and a default AAL test level in a water matrix.

5.4.2.1. Level B Method Validation: Same, Similar, or Slightly Different Matrix

Most qualified laboratories will have existing methods to analyze for the radionuclides of interest in the three most common matrices of water, air particulate filters and soil/sediment. Under method validation Level B, a method that has been previously validated for a different project and used for one matrix may be used for that same matrix or modified for use for a very similar matrix. An example of a slightly different matrix might be a method used for water samples having low dissolved solids, modified for water samples containing high dissolved solids. Level B requires the laboratory to conduct a method validation study for the radionuclide and matrix combination where three replicate samples from each of the three concentration levels are analyzed according to the method. Table 2 is to be used to determine the lower, middle, and upper testing levels for the replicate analyses. The test samples are internal PT samples prepared at the laboratory. In order to determine if a proposed method meets the project MQO requirements for the required method uncertainty, each internal PT sample result is compared with the method uncertainty acceptance criteria in Table 3. The acceptance criteria in Table 3 for Level B validation stipulate that, for each test sample analyzed, the measured value must be within ± 2.8 u_{MR} for test-level concentrations at or less than the AAL or ± 2.8 φ_{MR} for the test-level concentration above the AAL. These acceptance criteria apply to either established AALs stated in Appendix A or default AALs (Section 5.2). The values of u_{MR} and φ_{MR} for select radionuclide and matrix combinations are provided in Appendix A for established PAG AALs or in Table 4 for default AALs. The required method uncertainty values for the established PAG AALs are based on the *Radiological Laboratory Sample Analysis Guide for Incidents of National Significance* for the three basic matrices addressed: water, air and soil. The Table 4 values for u_{MR} are base on the default AALs stated in Table 1, and the φ_{MR} values are taken from the air and water editions of the *Radiological Laboratory Sample Analysis Guide for Incidents of National Significance*. For example, in Table 4 the u_{MR} value of 5.2 pCi/sample for specific alpha- emitting nuclides in a water matrix (column two, row two) is calculated by multiplying the relative required method uncertainty (φ_{MR}) of 0.13 (column three, row two) for this radionuclide and matrix combination by the default AAL in Table 1 (column three, row two) of 40 pCi/sample value. The φ_{MR} values used in Table 4 are 0.30 for screening measurements and 0.13 for specific radionuclide analyses. The u_{MR} values listed are in units of pCi per sample, assuming the aliquant sample sizes that will be used in the method validation process given in Table 1.

5.4.2.2. Level C Method Validation: New Application of an Existing Method to a Different Matrix

When a laboratory has a validated method for a radionuclide in one matrix but not others, the method may require modification to accommodate a completely different media/matrix (water versus soil) or a very different similar matrix (two soils of different physicochemical compositions). The degree of adaptation or modification needed will vary according to chemical and physical differences in the two matrices. Because of the extent of these differences, a laboratory may choose to validate the method for the radionuclide and matrix combination through method validation Level C. Level C method validation requires the laboratory to conduct a method validation study wherein five replicate samples from each of the three concentration levels are analyzed according to the method. Table 2 is to be used to determine the lower, mid and upper testing levels for the replicate analyses. The test samples are internal PT samples prepared at the laboratory. In order to determine if a proposed method meets the project MQO requirements for the required method uncertainty, each internal PT sample result is compared with the method uncertainty acceptance criteria of Table 3. The acceptance criteria stated in Table 3 for Level C for new applications stipulate that, for each test sample analyzed, the measured value must be within ±2.9 u_{MR} for test level concentrations at or less than the AAL or ± 2.9 φ_{MR} for the test level concentration above the AAL. These acceptance criteria apply to either established PAG AALs stated in Appendix A or default AALs (see Section 5.2). The values of u_{MR} and φ_{MR} for select radionuclide and matrix combinations are provided in Appendix A for established PAG AALs or in Table 4 for default AALs.

Table 4. Required Method Uncertainty (u_{MR} and φ_{MR}) Values for Default AAL Test Levels

Radionuclide [1]	Water [2,6] u_{MR} (pCi/ sample)	φ_{MR}	Air Sampling Media/Swipes [3,6] u_{MR} (pCi/ sample)	φ_{MR}	Solids [5,6] u_{MR} (pCi/ sample)	φ_{MR}
Gross α Screen	0.60	0.30	6.7	0.30	TBD	TBD
Specific Alpha-Emitting [4] Nuclides - based on ^{241}Am	5.2	0.13	0.73	0.13	TBD	TBD
Gross β Screen	3.6	0.30	580	0.30	TBD	TBD
Specific Beta-Emitting [4] Nuclides - based on ^{90}Sr	31	0.13	63	0.13	TBD	TBD
Gamma Screen	9.9	0.30	2,500	0.30	TBD	TBD
Specific Gamma-Emitting [4] Nuclides - based on ^{60}Co	86	0.13	270	0.13	TBD	TBD

Notes:
[1] For each radionuclide category, the absolute required method uncertainty (u_{MR}) is applied to the lower and mid test levels. The relative required method uncertainty (φ_{MR}) is to be used for the upper test level.
[2] Required method uncertainty values for water correspond to the 100-mrem dose-derived concentration values from Scenario 1 of the *Radiological Laboratory Sample Analysis Guide for Incidents of National Significance– Radionuclides in Water* (EPA 2008).
[3] Required method uncertainty values for air sampling media correspond to the 10^{-4} risk-based derived air concentration values from Scenario 1 of the *Radiological Laboratory Sample Analysis Guide for Incidents of National Significance– Radionuclides in Air* (EPA 2009a).
[4] The default values stated in the specific emitting nuclide rows apply to all radionuclides in the designated emission category. The reference to a radionuclide is presented only as information indicating the basis for the specific emission category.
[5] TBD: To be determined pending development of *Radiological Laboratory Sample Analysis Guide for Incidents of National Significance–Radionuclides in Soils* (In preparation).
[6] Table values have been rounded after calculations.

5.4.2.3. Level D Method Validation: Adapted or Newly Developed Methods, Including Rapid Methods

In some cases, a laboratory may not have a method for a certain radionuclide and matrix combination. For such situations, the laboratory may either develop a new method internally or adapt a method from the literature. In this case, the new method should undergo general method validation first and then incident response method validation. A laboratory would validate the new method for the radionuclide and matrix combination through method validation Level D or E (Section 5.4.2.4). Level D method validation requires the laboratory to conduct a method validation study wherein seven replicate samples from each of the three concentration levels are analyzed according to the method. Table 2 is to be used to determine the lower, mid and upper testing levels for the replicate analyses. For validation Level D, the test samples are internal PT samples prepared at the laboratory. In order to determine if a proposed method meets the project MQO requirements for the required method uncertainty, each internal PT sample result is compared with the method uncertainty acceptance criteria in Table 3. The acceptance criteria stated in Table 3 for Level D validation stipulate that, for each test sample analyzed, the measured value must be within ± 3.0 u_{MR} for test level concentrations at or less than the AAL or ± 3.0 φ_{MR} for the test level concentration above the AAL. These acceptance criteria apply to either established PAG AALs stated in Appendix A or default AALs (Section 5.2). The values of u_{MR} and φ_{MR} for select radionuclide and matrix combinations are provided in Appendix A for established AALs or in Table 4 for default AALs.

5.4.2.4. Level E Method Validation: Adapted or Newly Developed Methods, Including Rapid Methods, Using Method Validation Reference Materials

Methods developed by the laboratory or adapted from the literature that have undergone general method validation but not project method validation for an incident response are to be validated according to Levels D or E of Table 3. Both of these method validation levels have the same number of required test sample replicates and validation acceptance criteria. However, validation Level E is used when the sample matrix under consideration is unique and the IC determines that the method should be validated using the same matrix as the expected sample matrix. In this case, special method validation reference material (MVRM) would be used in the method validation process. The use of MVRM may be important for unique non-potable water matrices, soils or sediments, and manufactured type sample matrices.

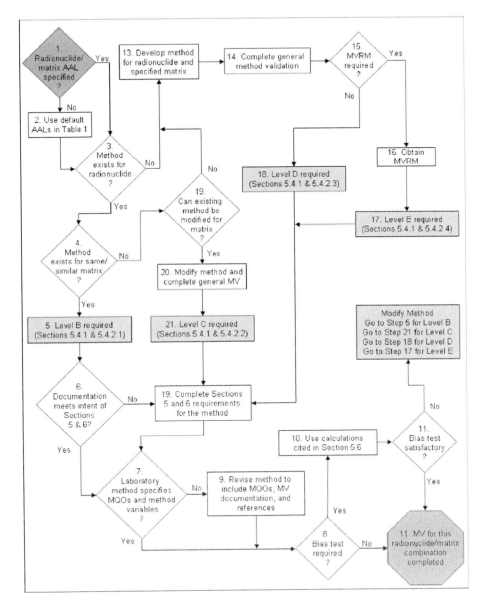

Figure 1. Method Validation Process for the Required Method Uncertainty MQO.

Level E method validation requires the laboratory to conduct a method validation study wherein seven replicate samples from each of the three concentration levels are analyzed according to the method. Table 2 is to be used to determine the lower, mid and upper testing levels for the replicate analyses. The test samples are external MVRM samples prepared for the laboratory. In order to determine if a proposed method meets the project MQO requirements for the required method uncertainty, each MVRM sample result is compared with the method uncertainty acceptance criteria of Table 3. The acceptance criteria stated in Table 3 for Level E validation stipulate that, for each test sample analyzed, the measured value must be within ±3.0 u_{MR} for test level concentrations at or less than the AAL or ± 3.0 φ_{MR} for the test level concentration above the AAL. These acceptance criteria apply to either established PAG AALs stated in Appendix A or default AALs (See Section 5.2). The values of u_{MR} and φ_{MR} for select radionuclide and matrix combinations are provided in Appendix A for established AALs or in Table 4 for default AALs.

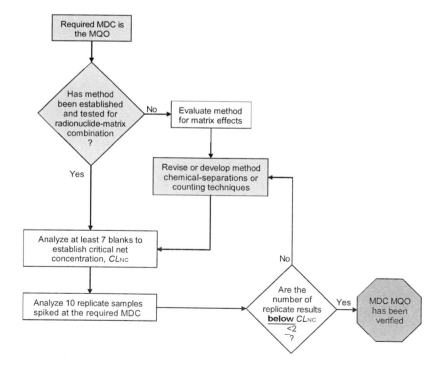

Figure 2. Validation Process for Verifying the Required MDC MQO.

Figure 1 (page 20) identifies the general approach to the method validation path that needs to be taken for a specific combination of radionuclide and matrix. Laboratories should use this chart to see where they are in the method validation process. It may be helpful for laboratories to create a similar flowchart in their method validation documentation to assist reviewers, auditors, and training personnel in recognizing the thought process used by the laboratory to validate methods.

5.5. Verification of Required Detection Limit (MDC) Specification

This section provides specifications for the method validation process to verify the required MDC specification MQO. Figure 2 (page 20) illustrates the process used to verify the required MDC for a method. The specifications presented are separate requirements with respect to the project method validation process for the required method uncertainty MQO. General method validation requirements are to be met prior to the initiation of this verification process. The specifications given in this section are distinct from those given for the required method uncertainty MQO method-validation process.

5.5.1. Calculation of the Critical Net Concentration

The critical net concentration shall be calculated for the required MDC method validation process. The calculation of the critical net concentration for the method is based on the analytical results of the blank matrix samples used in the MDC validation process. A minimum of seven blank samples is required. To ensure testing for sufficient method specificity, the matrix blanks should contain the anticipated concentration levels of chemical interferences and the potential interfering radionuclides (naturally occurring and incident response-related).

The critical net concentration (CL_{NC}), with a Type I error probability of $\alpha = 0.05$, is calculated using the following equation (consistent with MARLAP, Chapter 20, Equation 20.35):

$$CL_{NC} \text{ (pCi/unit)} = t_{1-\alpha}(n-1) \times s_{Blanks} \qquad (1)$$

where s_{Blanks} is the standard deviation of the n blank-sample net results (corrected for instrument background) in radionuclide concentration units of pCi/sample, and $t_{1-\alpha}(n-1)$ is the $(1-\alpha)$-quantile of the t-distribution with $n-1$

degrees of freedom (see MARLAP Table G.2 in Appendix G). Although the Type I error rate of 0.05 is routinely used and accepted, it is possible that other error rates may be used in incident response situations.

For seven (minimum) blank results (six degrees of freedom) and a Type I error probability of 0.05, Equation (1) reduces to:

$$CL_{NC} \text{ (pCi/unit)} = 1.94 \times s_{Blanks} \qquad (2)$$

If the number of blank samples is different than the minimum value of seven, refer to MARLAP Chapter 20, Attachment 20A for appropriate guidance. Care must be taken to ensure that all samples and blanks are analyzed under conditions that are typical of those used for routine analyses using the same sample weight or volume and with the same instruments with representative counting efficiencies and background levels. The calculated critical net concentration will be used in the verification process to determine if a method is capable of meeting the required MDC specification as described in Section 5.5.2.

5.5.2. Testing for the Required MDC

When a required MDC specification for a radionuclide and matrix combination is given as an MQO rather than the required method uncertainty, the method should be validated by verifying that the method can meet the required MDC. As noted in Table 5, method validation for the required MDC specifies that ten replicate samples, each spiked at the required MDC, should be analyzed and evaluated. In addition, the results of at least seven blank samples are used to determine the critical net concentration of the method (Section 5.5.1). The ten replicate spiked samples and seven blanks should contain the chemical species and potential interfering radionuclides which are reasonably expected to be present in an actual sample. To ensure the testing for sufficient method specificity, the expected concentration levels of the chemical species and potential interfering radionuclides should be used during testing. Figure 2 (page 20) provides an overview of the method validation process for verifying the required MDC MQO.

Testing for the required MDC verification is based on the null hypothesis that the true MDC for the method is at or below the required MDC. If the true MDC of the method has been calculated properly and is equal to or less than the required MDC, the probability of failing to detect the radionuclide at or above the critical net concentration is at most β. For project method validation related to incident response, β is assumed to be 0.05. The number of "non-detects" (sample results below the critical net concentration) for a set of n samples spiked at the required MDC is assumed to have a binomial distribution with parameters β and n. For a set of ten samples spiked at the required MDC, the number of Y sample results expected to be below the critical net concentration is not more than two (2) for a β of 0.05. If Y is greater than two, the null hypothesis is rejected.

The following protocol should be used to verify a method's capability to meet the required method MDC for each radionuclide (including gross screening)-matrix combination:

1. Analyze a minimum of seven blank samples (representing the matrix of interest) for the radionuclide under consideration.
2. From the blank sample net results, calculate the estimated *Critical Net Concentration* (Section 5.5.1), CL_{NC}.
3. Analyze ten replicate samples (representing the matrix of interest) spiked at the required MDC for the radionuclide under consideration.
4. From the results of the ten replicate samples spiked at the required MDC, determine the number (Y) of sample results at or below the estimated *Critical Net Concentration*.
5. If $Y \leq 2$, the method evaluated at the required MDC passes the test for the required MDC specification.
6. If $Y > 2$, the method evaluated at the required MDC fails the test for the required MDC specification.

Appendix C provides an example for testing a method's capability to meet a required MDC specification.

5.6. Method Bias Tests

In order to provide quality data, a method should not have a significant bias. Depending on the radiological incident, acceptable absolute and relative bias criteria for a method may be specified by the IC. Since the degree of acceptability of method bias depends on many parameters and circumstances, specific acceptance criteria for method bias have not been included for this method validation process. However, because the acceptance criteria for method uncertainty and required MDC verification will not tolerate a significant method bias or measurement uncertainty, acceptable method bias is indirectly evaluated when evaluating method uncertainty and the required

MDC. Appendix D provides an example of the effect of bias on the probability of failing the required method uncertainty validation acceptance criteria for method validation Level D.

Method bias is initially evaluated during method development, general and project method validation processes, and then continuously during the processing of incident response samples using batch QC samples (MARLAP Chapter 7). Tests for absolute and relative biases shall be made for the method validation level specified by the IC. The absolute bias shall be evaluated using the blank sample results and the relative bias evaluated for each test level (lower, mid and upper) using the results of the replicates.

When there is a significant absolute or relative bias, the probability of failing the required method uncertainty acceptance criteria of Section 5.4.2 may become significant depending on the magnitude of the actual method uncertainty. Appendix D provides an example of the probability of failing method validation Level D for three actual method uncertainty values (as compared to the required method uncertainty) as a function of relative bias up to 20%. In general, to avoid failure to meet the method validation acceptance criteria, it is best to have an actual method uncertainty at the AAL that is a fraction of the required method uncertainty.

The following equations, taken from MARLAP Chapter 6 (Attachment 6A) and other statistical references, are used to test for absolute and relative biases.

Table 5. Method Validation Requirements Applicable to Required MDC Verification

Method Characteristic	Application	Sample Type	Acceptance Criterion	Levels (Concentrations)	Replicates	# of Analyses
Detection Capability	Required MDC Specification	Internal PT	Number of Sample Results Below Critical Net Concentration Value ≤ 2	Single Concentration at the Required MDC Value	10	10

Note: At least seven blank samples should be analyzed to estimate the critical net concentration as part of the required MDC verification.

5.6.1. Absolute Bias Testing

The protocol for testing for absolute bias is the following:

1. Calculate the mean $(\overline{X})$ for "N" (at least seven) blank sample net results using Equation 3.

$$\overline{X} = \frac{1}{N}\sum_{i=1}^{N} X_i \qquad (3)$$

where N should be at least seven blank sample results.

2. Calculate the experimental standard deviation (s_x) of the same results[5] using Equation 4.

$$S_x = \sqrt{\frac{1}{N-1}\sum_{i=1}^{N}(X_i - \overline{X})^2} \qquad (4)$$

3. Use Equation 5 to calculate the |T| value:

$$|T| = \frac{|\overline{X}|}{s_x/\sqrt{N}} \qquad (5)$$

4. An absolute bias in the measurement process is indicated if

$$|T| > t_{1-\alpha/2}\,(N-1) \qquad (6)$$

where, $t_{1-\alpha/2}\,(N-1)$ represents the $(1 - \alpha/2)$-quantile of the t-distribution with $N-1$ degrees of freedom. For seven blanks, an absolute bias is identified at a significance level of 0.05, when $|T| > 2.447$.

5.6.2. Relative Bias Testing

5.6.2.1. Test Level Samples with Same Known Value

When the samples for a test level have the same concentration (e.g., water) or activity, the protocol for testing relative bias for each method validation test level is the following:

1. Calculate the mean ($\overline{X}$) and estimated standard deviation (s_X) of the replicate results for each method validation test level using Equations 3 and 4, respectively.
2. Use Equation 7 to calculate the |T| value

$$|T| = \frac{|\overline{X} - K|}{\sqrt{s_X^2/N + u^2(K)}} \qquad (7)$$

where:
- $\overline{X}$ is the average measured value
- s_X is the experimental standard deviation of the measured values
- N is the number of replicates
- K is the reference value
- $u(K)$ is the standard uncertainty of the reference value

3. A relative bias in the measurement process is indicated if

$$|T| > t_{1-\alpha/2}(v_{\text{eff}}) \qquad (8)$$

The number of *effective degrees of freedom* for the *T* statistic is calculated as follows:

$$v_{\text{eff}} = (N-1)\left(1 + \frac{u^2(K)}{s_X^2/N}\right)^2 \qquad (9)$$

v_{eff} as calculated by the equation generally is not an integer so v_{eff} should be truncated (rounded down) to an integer. Then, given the significance level of 0.05, the critical value for is defined to be $t_{1-\alpha/2}(v_{\text{eff}})$, the (1− α/2)-quantile of the *t*-distribution with v_{eff} degrees of freedom (see MARLAP Appendix G, Table G.2).

5.6.2.2. Test Level Samples with Slightly Different Known Values

When the PT samples for a test level have slightly different concentrations or activities (e.g., independently prepared[6] water samples, air filters, or swipes), the following protocol (paired *t*-test) for testing relative bias for each method validation test level is:

1. Calculate the average difference ($\overline{D}$) between the measured value and the known spiked value using Equation 10:

$$\overline{D} = \frac{1}{N}\sum_{i=1}^{N}(X_i - K_i) \tag{10}$$

where
X_i is the measured value for the ith sample at a particular test level
K_i is the known value for the same sample
N is the number of samples at that test level

2. Calculate the standard deviation of the differences, S_D, as:

$$S_D = \sqrt{\frac{1}{N-1}\sum_{i=1}^{N}(D_i - \overline{D})^2} \tag{11}$$

where $D_i = X_i - K_i$.

3. Calculate the absolute value of the t statistic as:

$$|T| = \frac{|\overline{D}|}{S_D/\sqrt{N}} \tag{12}$$

4. A relative bias in the measurement process for a test level is indicated if

$$|T| > t_{1-\alpha/2}(N-1) \tag{13}$$

6.0. Method Validation Documentation

The information and data to be retained should be specified in the method validation plan for each radionuclide and matrix combination. When the laboratory conducts project method validation for incident response applications, the detailed analytical method and all records, laboratory workbooks, and matrix spike data used to validate the analytical method should be retained on file and be retrievable for a specified length of time after the method has been discontinued. Data evaluations such as comparison of individual results to the validation acceptance criteria and absolute bias in blanks and, when available, method precision and bias, should be part of the data validation package retained as part of the documentation related to the laboratory's quality system. In addition, for each radionuclide and matrix combination, a synoptic method validation report containing the analytical method identification, method validation acceptance criteria, test levels, validation results and a method acceptability decision should be generated and retained.

REFERENCES

[1] *American National Standards Institute (ANSI) N42.23. Reapproved 2003*. Measurement and Associated Instrumentation Quality Assurance for Radioassay Laboratories.

[2] U.S. Environmental Protection Agency (EPA). (2006). *Validation and Peer Review of U.S. Environmental Protection Agency Radiochemical Methods of Analysis*. FEM Document Number 2006-01, November 8.

[3] U.S. Environmental Protection Agency (EPA). (2008). *Radiological Laboratory Sample Analysis Guide for Incidents of National Significance–Radionuclides in Water*. Revision 0. Office of Air and Radiation, Washington, DC. EPA 402-R-07-007, January. Available at: www.epa.gov/ narel/recent_info.html.

[4] U.S. Environmental Protection Agency (EPA). (2009a). *Radiological Laboratory Sample Analysis Guide for Incidents of National Significance–Radionuclides in Air*. Revision 0. Office of Air and Radiation, Washington, DC. EPA 402-R-09-007, June. Available at: www.epa.gov/narel/recent_ info.html.

[5] U.S. Environmental Protection Agency (EPA). (2009b). *Radiological Laboratory Sample Screening Analysis Guide for Incidents of National Significance*. Revision 0. Office of Air and Radiation, Washington, DC. EPA 402-R-09-008, June. Available at: www.epa.gov/narel/recent_info.html.

[6] *EURACHEM Guide*. (1998). *The Fitness for Purpose of Analytical Methods, A Laboratory Guide to Method Validation and Related Topics*. Available at: http://www.eurachem.org/.

[7] *International Standard ISO/IEC 17025. (2005). General Requirements for the Competence of Testing and Calibration Laboratories*. ISO,

[8]   International Organization for Standardization (ISO). (1993). *International Vocabulary of Basic and General Terms in Metrology.* 2nd Edition. ISO, Geneva, Switzerland.
[9]   *Multi-Agency Radiological Laboratory Analytical Protocols Manual* (MARLAP). (2004). EPA 402- B-04-001A, July. Volume I, Chapters 6, 7, 20, Glossary; Volume II and Volume III, Appendix G. Available at: www.epa.gov/radiation/marlap.
[10] *Multi-Agency Radiation Survey and Site Investigation Manual* (MARSSIM), *Revision 1.* (2000). NUREG-1575 Rev 1, EPA 402-R-97-016 Rev1, DOE/EH-0624 Rev1. August. Available from www.epa.gov/radiation/marrsim.

APPENDIX A: TABLES SUMMARIZING THE DERIVED RADIONUCLIDE CONCENTRATIONS AND REQUIRED METHOD UNCERTAINTIES CORRESPONDING TO PAGS OR RISKS FOR THE WATER, AIR, AND SOIL MATRICES

Table A1. Alpha-Emitting Radionuclide Concentrations and Required Method Uncertainties in Water Corresponding to 500- and 100-mrem AAL Derived Water Concentrations (DWCs)

Radionuclide	pCi/L					
	500 mrem		100 mrem			
	AAL DWC[1][2]	Screening Methods Required Method Uncertainty (u_{MR}) [6]	AAL DWC[1][2][3]	Screening Methods Required Method Uncertainty (u_{MR}) [6]	Nuclide-Specific Required Method Uncertainty (u_{MR}) [6]
---	---	---	---	---	---
Gross α Screen [5]	2.0×10^3	610	400	120	—
Am-241	2.0×10^3	610	400	120	50
Cm-242	1.4×10^4	4.3×10^3	2.8×10^3	850	350
Cm-243	2.5×10^3	760	500	150	63
Cm-244	2.9×10^3	880	580	180	73
Np-237	3.9×10^3	1.2×10^3	780	240	98
Po-210	130	40	26	7.9	3.3
Pu-238	1.8×10^3	550	360	110	45
Pu-239	1.7×10^3	520	340	100	43

Table A1. (Continued)

Radionuclide	pCi/L				
	500 mrem		100 mrem		
	AAL DWC[1][2]	Screening Methods Required Method Uncertainty (u_{MR}) [6]	AAL DWC[1][2][3]	Screening Methods Required Method Uncertainty (u_{MR}) [6]	Nuclide-Specific Required Method Uncertainty (u_{MR}) [6]
Pu-240	1.7×10^3	520	340	100	43
Ra-226 [4]	910	280	180	55	23
Th-228 [4]	2.6×10^3	790	520	160	65
Th-230	1.8×10^3	550	360	110	45
Th-232	1.6×10^3	490	320	97	40
U-234	6.3×10^3	1.9×10^3	1.3×10^3	400	160
U-235	6.6×10^3	2.0×10^3	1.3×10^3	400	160
U-238	7.0×10^3	2.1×10^3	1.4×10^3	430	180

Notes:

[1] Values are based on the dose conversion factors in Federal Guidance Report No.13, CD Supplement, 5- year-old child and the 50th percentile of water consumption.

[2] 365-day intake.

[3] Values obtained by dividing 500-mrem DWC values by 5.

[4] Includes the dose from the decay products originating from the ^{226}Ra or ^{228}Th in the body.

[5] Values for gross alpha screening are based on ^{241}Am.

[6] The required relative method uncertainty (φ_{MR}) for values greater than the AALs in this table is obtained by dividing the u_{MR} value by the corresponding AAL.

Table A2. Beta/Gamma-Emitting Radionuclide Concentrations in Water and Required Method Uncertainties Corresponding to 500- and 100-mrem AAL Derived Water Concentrations (DWCs)

Radionuclide	pCi/L				Nuclide-Specific Required Method Uncertainty (u_{MR})[5]
	500 mrem		100 mrem		
	AAL DWC [1][2]	Screening Methods Required Method Uncertainty (u_{MR})[5]	AAL DWC [1][2][3]	Screening Methods Required Method Uncertainty (u_{MR})[5]	
Gross β/γ Screen[4]	5.8×10^4	1.8×10^4	1.2×10^4	3.6×10^3	—
Ac-227DP[6]	1.1×10^3	330	220	67	28
Ce-141	2.2×10^5	6.7×10^4	4.4×10^4	1.3×10^4	5.5×10^3
Ce-144	2.9×10^4	8.8×10^3	5.8×10^3	1.8×10^3	730
Co-57	6.3×10^5	1.9×10^5	1.3×10^5	4.0×10^4	1.6×10^4
Co-60	3.3×10^4	1.0×10^4	6.6×10^3	2.0×10^3	830
Cs-134	4.3×10^4	1.3×10^4	8.6×10^3	2.6×10^3	1.1×10^3
Cs-137	5.8×10^4	1.8×10^4	1.2×10^4	3.6×10^3	1.5×10^3
H-3	7.7×10^6	2.3×10^6	1.5×10^6	4.6×10^5	1.9×10^5
I-125	1.3×10^4	4.0×10^3	2.6×10^3	790	320
I-129	3.3×10^3	1.0×10^3	660	200	83
I-131	5.4×10^3	1.6×10^3	1.1×10^3	330	140
Ir-192	1.2×10^5	3.6×10^4	2.4×10^4	7.3×10^3	3.0×10^3
Mo-99	3.2×10^5	9.7×10^4	6.4×10^4	1.9×10^4	8.1×10^3
P-32	5.9×10^4	1.8×10^4	1.2×10^4	3.6×10^3	1.5×10^3
Pd-103	7.8×10^5	2.4×10^5	1.6×10^5	4.9×10^4	2.0×10^4
Pu-241	1.0×10^5	3.0×10^4	2.0×10^4	6.1×10^3	2.5×10^3
Ra-228[6]	160	49	32	9.7	4.0
Ru-103	2.3×10^5	7.0×10^4	4.6×10^4	1.4×10^4	5.8×10^3
Ru-106	2.2×10^4	6.7×10^3	4.4×10^3	1.3×10^3	550
Se-75	6.7×10^4	2.0×10^4	1.3×10^4	4.0×10^3	1.6×10^3
Sr-89	6.3×10^4	1.9×10^4	1.3×10^4	4.0×10^3	1.6×10^3
Sr-90	1.2×10^4	3.6×10^3	2.4×10^3	730	300
Tc-99	2.4×10^5	7.3×10^4	4.8×10^4	1.5×10^4	6.0×10^3

Notes:
[1]. Values are based on the dose conversion factors in Federal Guidance Report No.13, CD Supplement, 5-year-old child and the 50th percentile of water consumption.
[2] 365-day intake.
[3]. Values obtained by dividing 500-mrem DWC values by 5.
[4] Gross beta screening values are based on ^{137}Cs.
[5] The required relative method uncertainty (φ_{MR}) for values greater than the AALs is obtained by dividing the u_{MR} value in this table by the corresponding AAL value.
[6] Includes the dose from the decay products originating from the ^{228}Ra or ^{227}Ac in the body.

Table A3. Alpha-Emitting Radionuclide Concentrations in Air and Required Method Uncertainties Corresponding to 2-rem and 500-mrem AAL Derived Air Concentrations (DACs)

Radionuclide	pCi/m³						
	2 rem				500 mrem		
	AAL DAC [1]	Screening Method Required Method Uncertainty (u_{MR}) [3]	Nuclide-Specific Required Method Uncertainty (u_{MR}) [3]	AAL DAC[1]	Screening Method Required Method Uncertainty (u_{MR}) [3]	Nuclide-Specific Required Method Uncertainty (u_{MR}) [3]
Gross α Screen[4]	0.70	0.21	—	0.17	0.052	—
Am-241	0.70	0.21	0.088	0.17	0.052	0.021
Cm-242	11	3.3	1.4	2.8	0.85	0.35
Cm-243	0.97	0.29	0.12	0.24	0.073	0.030
Cm-244	1.2	0.36	0.15	0.29	0.088	0.037
Np-237	1.3	0.40	0.16	0.34	0.10	0.043
Po-210	16	4.9	2.0	3.9	1.2	0.49
Pu-238	0.62	0.19	0.081	0.15	0.046	0.020
Pu-239	0.56	0.17	0.071	0.14	0.043	0.018
Pu-240	0.56	0.17	0.071	0.14	0.043	0.018
Ra-226 [2]	7.0	2.1	0.88	1.8	0.55	0.23
Th-228 [2]	1.7	0.52	0.21	0.42	0.13	0.053
Th-230	0.66	0.20	0.083	0.17	0.052	0.021
Th-232	0.61	0.19	0.077	0.15	0.046	0.019
U-234	7.1	2.2	0.89	1.8	0.55	0.23
U-235	7.9	2.4	0.99	2.0	0.61	0.25
U-238	8.3	2.5	1.0	2.1	0.64	0.26

Notes:

[1] Morbidity for long-term inhalation. Child as receptor. Value corresponds to solubility class having lowest value.

[2] Includes the dose from the decay products originating from the ^{226}Ra or ^{228}Th in the body.

[3] Required method uncertainty values are based on a sampled aerosol volume of 68m³ at the 2 rem or 500- mrem DAC. The required relative method uncertainty (φ_{MR}) for values greater than the AALs in this table is obtained by dividing the u_{MR} value in this table by the corresponding AAL value.

[4] The gross α screening values are not related to a specific radionuclide.

Table A4. Beta/Gamma-Emitting Radionuclide Concentrations in Air and Required Method Uncertainties Corresponding to 2-rem and 500-mrem AAL Derived Air Concentrations (DACs)

Radionuclide	pCi/m³					
	2 rem			500 mrem		
	AAL DAC [1,4]	Screening Method Required Method Uncertainty (u_{MR}) [3]	Nuclide-Specific Required Method Uncertainty (u_{MR}) [3]	AAL DAC [1,4]	Screening Method Required Method Uncertainty (u_{MR}) [3]	Nuclide-Specific Required Method Uncertainty (u_{MR}) [3]
Gross β Screen [5]	420	130	—	110	33	—
Ac-227+DP [2]	0.43	0.13	0.054	0.11	0.033	0.014
Ce-141	1.8×10⁴	5.5×10³	2.3×10³	4.5×10³	1.4×10³	570
Ce-144	1.3×10³	400	160	320	97	40
Co-57	6.7×10⁴	2.0×10⁴	8.4×10³	1.7×10⁴	5.2×10³	2.1×10³
Co-60	2.2×10³	670	280	540	170	69
Cs-134	3.3×10³	1.0×10³	420	820	250	100
Cs-137	1.7×10³	520	210	430	130	54
H-3	2.6×10⁵	7.9×10⁴	3.3×10⁴	6.4×10⁴	1.9×10⁴	8.1×10³
I-125 [6]	1.3×10⁴	4.0×10³	1.6×10³	3.2×10³	970	400
I-129 [6]	1.9×10³	580	240	470	140	59
I-131 [6]	9.1×10³	2.8×10³	1.1×10³	2.3×10³	700	290
Ir-192	1.0×10⁴	3.0×10³	1.3×10³	2.5×10³	760	310
Mo-99	6.8×10⁴	2.1×10⁴	8.6×10³	1.7×10⁴	5.2×10³	2.1×10³
P-32	1.7×10⁴	5.2×10³	2.1×10³	4.3×10³	1.3×10³	540
Pd-103	1.5×10⁵	4.6×10⁴	1.9×10⁴	3.8×10⁴	1.2×10⁴	4.8×10³
Pu-241	29	8.8	3.7	7.3	2.2	0.92
Ra-228 [2]	4.2	1.3	0.53	1.0	0.30	0.13
Ru-103	2.3×10⁴	7.0×10³	2.9×10³	5.7×10³	1.7×10³	720
Ru-106	1.0×10³	300	130	250	76	31
Se-75	5.0×10⁴	1.5×10⁴	6.3×10³	1.3×10⁴	4.0×10³	1.6×10³
Sr-89	8.4×10³	2.6×10³	1.1×10³	2.1×10³	640	260
Sr-90	420	130	53	110	33	14
Tc-99	5.0×10³	1.5×10³	630	1.3×10³	400	160

Notes:

[1] Derived air concentration yielding stated committed effective dose assuming a 365-day year. Child as receptor. Value corresponds to solubility class having lowest value.

[2] Includes the dose from the decay products originating from the ^{228}Ra or ^{227}Ac in the body. DP refers to "decay products."

[3] Required method uncertainty values are based on a sampled aerosol volume of 68 m³ at the 2 rem or 500-mrem DAC. The required relative method uncertainty

(φ_{MR}) for values greater than the AALs in this table is obtained by dividing the u_{MR} value in this table by the corresponding AAL value.
[4] All nuclides can be collected on a fibrous or membrane air filter media except ^{3}H, ^{125}I, ^{129}I, and ^{131}I in the vapor states.
[5] Gross beta screening values are based on ^{90}Sr.
[6] These values are based on the vapor plus particulate dose rate.

Table A5. Alpha-Emitting Radionuclide Concentrations in Air and Required Method Uncertainties Corresponding to AAL Derived Air Concentrations (DACs)

Radionuclide	pCi/m³			
	10^{-4} Risk AAL DAC[1]	10^{-4} Risk AAL Required Method Uncertainty (u_{MR})[3]	10^{-6} Risk AAL DAC[1]	10^{-6} Risk AAL Required Method Uncertainty (u_{MR})[3]
Gross α Screen [4]	0.33	0.042	3.3×10^{-3}	4.2×10^{-4}
Am-241	0.33	0.042	3.3×10^{-3}	4.2×10^{-4}
Cm-242	0.62	0.078	6.2×10^{-3}	7.8×10^{-4}
Cm-243	0.34	0.043	3.4×10^{-3}	4.3×10^{-4}
Cm-244	0.35	0.044	3.5×10^{-3}	4.4×10^{-4}
Np-237	0.43	0.054	4.3×10^{-3}	5.4×10^{-4}
Po-210	0.86	0.11	8.6×10^{-3}	1.1×10^{-3}
Pu-238	0.24	0.030	2.4×10^{-3}	3.0×10^{-4}
Pu-239	0.22	0.028	2.2×10^{-3}	2.8×10^{-4}
Pu-240	0.22	0.028	2.2×10^{-3}	2.8×10^{-4}
Ra-226 [2]	0.44	0.055	4.4×10^{-3}	5.5×10^{-4}
Th-228 [2]	0.094	0.012	9.4×10^{-4}	1.2×10^{-4}
Th-230	0.36	0.045	3.6×10^{-3}	4.5×10^{-4}
Th-232	0.30	0.038	3.0×10^{-3}	3.8×10^{-4}
U-234	0.45	0.057	4.5×10^{-3}	5.7×10^{-4}
U-235	0.49	0.062	4.9×10^{-3}	6.2×10^{-4}
U-238	0.52	0.065	5.2×10^{-3}	6.5×10^{-4}

Notes:
[1] Morbidity for long-term inhalation. Value corresponds to solubility class having lowest value.
[2]. Includes the dose from the decay products originating from the ^{226}Ra or ^{228}Th in the body.
[3]. Required method uncertainty values are based on a sampled aerosol volume of 1,600 m³ at the 10^{-4} and 10^{-6} risk DACs, respectively. The required relative method uncertainty (φ_{MR}) for values greater than the AALs in the table is obtained by dividing the u_{MR} value by the corresponding AAL value.
[4] The gross α screening values are not related to a specific radionuclide.

Table A6. Beta/Gamma-Emitting Radionuclide Concentrations in Air and Required Method Uncertainties Corresponding to AAL-Derived Air Concentrations (DACs)

Radionuclide	pCi/m³			
	10^{-4} Risk AAL DAC [1,4]	10^{-4} Risk AAL Required Method Uncertainty (u_{MR}) [3]	10^{-6} Risk AAL DAC [1,4]	10^{-6} Risk AAL Required Method Uncertainty (u_{MR}) [3]
Gross β Screen (Sr-90)	29	3.8	0.29	0.038
Ac-227+DP [2]	0.083	0.010	8.3×10^{-4}	1.0×10^{-4}
Ce-141	920	120	9.2	1.2
Ce-144	69	8.7	0.69	0.087
Co-57	3.3×10^{3}	420	33	4.2
Co-60	120	15	1.2	0.15
Cs-134	180	23	1.8	0.23
Cs-137	110	14	1.1	0.14
H-3 Vapor	1.5×10^{4}	1.9×10^{3}	150	19
I-125	1.2×10^{3}	150	12	1.5
I-129	200	25	2	0.25
I-131	640	81	6.4	0.81
Ir-192	510	64	5.1	0.64
Mo-99	2.6×10^{3}	330	26	3.3
P-32	890	110	8.9	1.1
Pd-103	7.0×10^{3}	880	70	8.8
Pu-241	14	1.8	0.14	0.018
Ra-228 [2]	0.28	3.5×10^{-2}	2.8×10^{-3}	3.5×10^{-4}
Ru-103	1.2×10^{3}	150	12	1.5
Ru-106	56	7.1	0.56	0.071
Se-75	2.5×10^{3}	310	25	3.1
Sr-89	410	52	4.1	0.52
Sr-90	29	3.7	0.29	0.037
Tc-99	330	42	3.3	0.42

Notes:
[1]. Morbidity for long-term inhalation. Value corresponds to solubility class having lowest value.
[2]. Includes the dose from the decay products originating from the ^{228}Ra or ^{227}Ac in the body.
[3]. Required method uncertainty values are based on a sampled aerosol volume of 1,600 m³ at the 1 0^{-4} and 1 0^{-6} risk DAC, respectively. The required relative method uncertainty ($_{uMR}$) for values greater than the AALs in the table is obtained by dividing the uMR value by the corresponding AAL value.

All nuclides can be collected on a fibrous or membrane air filter media except ^{3}H, ^{125}I, ^{129}I, and ^{131}I in the vapor states.

Table A7. Alpha and Beta/Gamma-Emitting Radionuclide Concentrations in Soil and Required Method Uncertainties Corresponding to Derived Soil Concentrations

Table to be determined following publication of Radiological Laboratory Sample Analysis Guide for Incidents of National Significance–Radionuclides in Soil.

APPENDIX B: EXAMPLES OF THE METHOD VALIDATION PROCESS FOR REQUIRED METHOD UNCERTAINTY SPECIFICATIONS

Two examples are provided to demonstrate the method validation process when the MQO involves a required method uncertainty (u_{MR} or φ_{MR}) specification. The first example is when an Incident Commander (IC) specifies a required method uncertainty for a method and an AAL (PAG or risk-based derived radionuclide concentration) for a typical radionuclide and matrix combination as provided in Appendix A. The radionuclide and matrix combination for this first example is ^{241}Am in potable water. Values for the derived radionuclide concentration AAL and required method uncertainty were obtained from the *Radiological Laboratory Sample Analysis Guide for Incidents of National Significance–Radionuclides in Water* (EPA 2008). The three testing level concentrations were determined using the AAL concentration value and the lower, mid and upper test level multipliers in Table 2. Acceptable validation criteria are established in Table 3 for Validation Level D.

The second example is when an Incident Commander specifies a required method uncertainty for a method used to analyze a radionuclide or matrix not provided in Appendix A. For this example, the radionuclide and matrix combination is ^{241}Am in street runoff water. Values for the derived AALs and required method uncertainty were obtained from Tables 1 and 4, respectively. The three testing level concentrations were determined using the AAL concentration value and the lower, mid and upper test level multipliers in Table 2. Acceptable validation criteria are established in Table 3 for validation Level D.

Example 1. Method Validation for Am-241 in Potable Water; Established AALs

Nuclide: ^{241}Am
Matrix: Water
Analytical Action Level: 400 pCi/L (Appendix A, Table A1, 100 mrem)
Proposed Method: Radiochemistry with alpha spectrometry
Required Method Validation Level: D
Required Method Uncertainty[7]: 50 pCi/L at AAL or below; 13% above AAL
Acceptance Criteria (Table 3): Measured value within ±3 u_{MR} (± 150 pCi/L) of known value #AAL and ±3 φ_{MR} (± 39%) of known value > AAL.
Test levels (Table 2): Lower (0.5 × AAL = 200 pCi/L; Mid (AAL) = 400 pCi/L; Upper (3 × AAL) = 1,200 pCi/L

Data Evaluation

Table B1. Required Method Uncertainty for Am-241 in Potable Water

Test Sample	Lower Test Level Concentration 200 pCi/L		Mid Test Level Concentration[1] 400 pCi/L[2]		Upper Test Level Concentration 1,200 pCi/L	
	Acceptable Range: 50 to 350 pCi/L		Acceptable Range: 250 to 550 pCi/L		Acceptable Range: 732 to 1,670 pCi/L	
	Measured Value ± 1 CSU[3]	Acceptable Value (Y/N)	Measured Value ± 1 CSU[3]	Acceptable Value (Y/N)	Measured Value ± 1 CSU[3]	Acceptable Value (Y/N)
1	221 ± 27	Y	429 ± 40	Y	1,283 ± 87	Y
2	179 ± 24	Y	381 ± 37	Y	1,117 ± 78	Y
3	210 ± 26	Y	405 ± 39	Y	1,241 ± 85	Y
4	190 ± 25	Y	304 ± 32	Y	1,159 ± 80	Y
5	169 ± 25	Y	362 ± 36	Y	1,262 ± 86	Y
6	225 ± 27	Y	458 ± 42	Y	1,138 ± 79	Y
7	213 ± 26	Y	390 ± 38	Y	994 ± 72	Y

Notes:
[1] Mid test level is at the AAL (Table 2)
[2] AAL taken from Table 9A, *Radiological Laboratory Sample Analysis Guide for Incidents of National Significance– Radionuclides in Water* (EPA 2008).
[3] Approximate combined standard uncertainty for a 100-minute count on an alpha detector having a typical detector efficiency plus another 5% uncertainty for other method parameters at the action level. Counting time was estimated so that the required method uncertainty would be met at the AAL. All samples would be counted for the same length of time regardless of the test level.

Example 2. Method Validation for Am-241 in Street Runoff Water - Default AAL and Required Method Uncertainty
Nuclide: ^{241}Am
Matrix: Street runoff water
***Default AAL*:** 40 pCi/sample (Table 1, liquid, specific nuclide)
***Proposed Method*:** Radiochemistry with alpha spectrometry. Specific nuclide measurement.
***Required Method Validation Level*:** D, new matrix
***Required Method Uncertainty*:** 5.2 pCi/test sample at AAL or below; 13% above AAL (Table 4)
***Acceptance Criteria* (Table 2):** Measured Value within ±3 u_{MR} (±15.6 pCi/sample) of known value #AAL and ±3 nMR (± 39%) of known value > AAL
***Method Validation Test Levels* (Table 2):** Lower (0.5 × AAL) = 20 pCi/sample; Mid (AAL) = 40 pCi/sample; Upper (3 × AAL) = 120 pCi/sample

Data Evaluation

Table B2. Required Method Uncertainty for Am-241 in Street Runoff Water

Test Sample	Lower Test Level Concentration 20 pCi/sample Acceptable Range: 4.4 to 35.6 pCi Measured Value ± 1 CSU[3]	Acceptable Value (Y/N)	Mid Test Level Concentration[1] 40 pCi/sample[2] Acceptable Range: 24.4 to 55.6 pCi Measured Value ± 1 CSU[3]	Acceptable Value (Y/N)	Upper Test Level Concentration 120 pCi/sample Acceptable Range: 73.2 to 167 pCi Measured Value ± 1 CSU[3]	Acceptable Value (Y/N)
1	22.3 ± 2.3	Y	44.2 ± 3.6	Y	128.6 ± 8.0	Y
2	17.6 ± 2.0	Y	36.7 ± 3.2	Y	112.2 ± 7.2	Y
3	20.9 ± 2.2	Y	42.4 ± 3.5	Y	124.7 ± 7.8	Y
4	23.4 ± 2.4	Y	38.1 ± 3.2	Y	117.0 ± 7.4	Y
5	15.8 ± 1.9	Y	50.5 ± 3.9	Y	140.0 ± 8.6	Y
6	21.7 ± 2.2	Y	41.5 ± 3.4	Y	122.0 ± 7.7	Y
7	18.8 ± 2.1	Y	31.1 ± 2.8	Y	113.4 ± 7.2	Y

Notes:
[1] Mid test level is at the AAL, (Table 2)
[2] Table 1, liquid, specific nuclide
[3] Approximate combined standard uncertainty for a 15-minute count on an alpha detector having a typical detector efficiency plus another 5% uncertainty for other method parameters at the action level. Counting time was estimated so that the required method uncertainty would be met at the AAL. All samples would be

counted for the same length of time regardless of the test level. Sample volume ~ 100 mL.

APPENDIX C: EXAMPLE OF THE METHOD VALIDATION PROCESS FOR VERIFICATION OF THE REQUIRED MDC SPECIFICATION

Refer to Section 5.5.2 (page 21) for the protocol to follow for verifying that a method's MDC meets the required MDC specification.

Nuclide: ^{90}Sr
Matrix: Street runoff water
Required MDC = 2 pCi/L (MQO designated by Incident Commander)
Proposed Method: Radiochemistry with beta counting on gas proportional counter. Sample volume = 1 L, counting time = 240 minutes. Analytical result calculations to include detector efficiency, detector background (cpm) and ^{90}Y ingrowth factor.
Number of Blanks: 7
Number of Spiked Test Samples: 10
Testing Level: 2 pCi/L of ^{90}Sr

Calculations:
(a) ^{90}Sr concentration and associated combined standard uncertainty for the blanks and test samples.
(b) Critical Net Concentration = 1.94 × standard deviation of the seven blank results.
(c) Number (Y) of sample results at or below the estimated Critical Net Concentration

Table C1. Results of Blank Sample Analyses

Blank Number	Result (pCi/L)
1	−0.21 ± 0.44
2	0.10 ± 0.45
3	0.44 ± 0.46
4	0.82 ± 0.46
5	−0.40 ± 0.44
6	−0.75 ± 0.44
7	0.61 ± 0.46
Average	0.09
Standard Deviation of Results	0.57
Critical Net Concentration	1.11

Table C2. Results of MDC Test Sample Analyses; Test Concentration = 2.0 pCi/L

Test Sample Number	Result (pCi/L)	Result ≤ Critical Net Concentration (1.11 pCi/L)
1	2.57 ± 0.50	N
2	1.00 ± 0.47	Y
3	2.43 ± 0.50	N
4	1.57 ± 0.48	N
5	2.29 ± 0.50	N
6	1.71 ± 0.48	N
7	2.01 ± 0.49	N
8	3.14 ± 0.52	N
9	0.86 ± 0.46	Y
10	1.43 ± 0.48	N
Average	1.90	
Standard Deviation of Results	0.72	
Y - Number of Results ≤ Critical Net Concentration		2

Conclusion: The hypothesis that the true MDC for the method is at or below the required MDC cannot be rejected. Therefore, the method is assumed to be capable of meeting the required MDC specification.

Test: Does the number (Y) of sample results at or below the estimated *Critical Net Concentration* exceed 2?

- If $Y \leq 2$, the method tested at the required MDC passes the test for the required MDC specification.
- If $Y > 2$, the method tested at the required MDC fails the test for the required MDC specification.

APPENDIX D: EXAMPLE OF THE EFFECT OF BIAS ON THE PROBABILITY OF FAILING THE METHOD VALIDATION ACCEPTANCE CRITERIA FOR REQUIRED METHOD UNCERTAINTY

Suppose one is validating a method for water using Level D acceptance criteria, so tests should be made at three concentration levels with seven samples at each level. Consider that the action level is 100 pCi/L, and that this is one of the test levels. Also suppose that the required method uncertainty at 100 pCi/L is u_{MR} = 10 pCi/L, i.e. the relative required method uncertainty is φ_{MR} = 10%. The acceptance bounds are then 100 ± 30 pCi/L. For the purpose of this illustration, only a positive method bias will be considered (although the same effect would occur for negative biases).

In Figure D1, the area under the curves above 130 pCi/L is the probability that a single sample will fail validation for the given method uncertainty and bias. If the method just meets the criterion and there is no bias, the figure shows that the probability of an individual sample failing is very small (<0.01%). Now suppose there is a bias of +10%. What is the probability of failing a single sample? When the uncertainty is 10 pCi/L it is 2.28%. Of course, if the method uncertainty already exceeds the required 10%, (e.g., 12.5 pCi/L), this probability is even higher, 5.48%. However, if the method uncertainty is less than that required (e.g., 7.5 pCi/L or 5 pCi/L), then this probability becomes lower, 0.38% and 0.13%, respectively.

The probabilities above are for a single sample. At each level, seven samples must pass. Because there are three levels, there are 21 tests, and if the probability of failure for a single sample is F, then the overall probability of failure is about $1 - (1-F)^{21}$. If F is 2%, $1 - (1-F)^{21} \gg 35\%$.

Consider an example of method validation Level D for a water matrix. There are 3 test levels with 7 samples each, or 21 total samples:

- 0.5 ×AAL level = 50 pCi/L; u_{Req} = 10 pCi/L, φ_{Req} = 0.20
- AAL level   = 100 pCi/L; u_{Req} = 10 pCi/L, φ_{Req} = 0.10
- 3.0 ×AAL level = 300 pCi/L; u_{Req} = 30 pCi/L, φ_{Req} = 0.10

The acceptance criterion is that each measured value (for all 21 samples) must be within ±3.0 u_{Req} or ±3.0 φ_{Req} of the validation test activity. The probability of passing the acceptance criteria was calculated for four assumed relative method uncertainty values: 5%, 7.5%, 10%, and 12.5%. The range of biases evaluated included 0 to 20% at the AAL.

Figure D2 shows the overall probability of failing the Level D validation as a function of bias and relative method uncertainty. It is clear that if the method just meets the required relative method uncertainty, then there is not much room to accommodate bias. However, when the relative method uncertainty is 7.5%, about an equal amount of bias might be tolerated. If the relative method uncertainty is half that required, three times as much bias can exist without bringing the probability of failure over 5%.

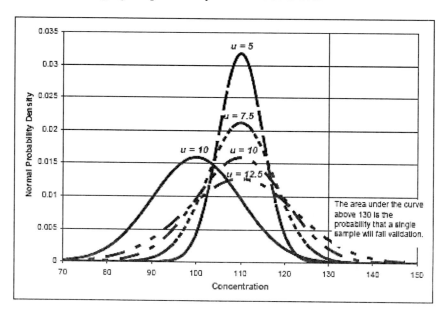

Figure D1. Probability of a validation sample failing at concentration 100 pCi/L with and without bias at various values of the method standard uncertainty.

Method Validation Guide for Qualifying Methods Used ... 55

- When the actual relative method uncertainty is one-half or less than the relative required method uncertainty, biases up to 15% may be tolerated without substantially increasing the probability of failing the acceptance criteria.
- When the actual relative method uncertainty is equal to the relative required method uncertainty, it is best not to have a bias in order to maintain a reasonable probability of passing the acceptance criteria.
- When the actual relative method uncertainty is greater than the relative required method uncertainty, the probability of failing the acceptance criteria is extremely high, regardless of the magnitude of the bias.

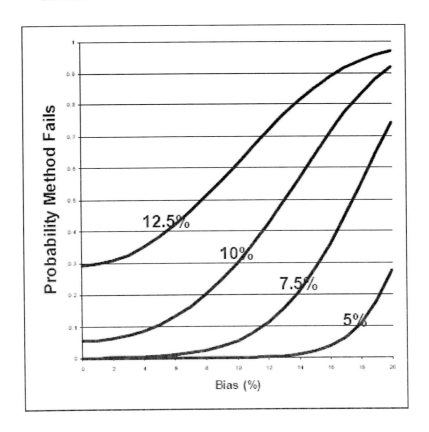

Figure D2. Level D validation (21 samples) failing at test level as a function of relative method bias for relative method uncertainties of 5%, 7.5%, 10%, and 12.5%.

Detecting Bias

Testing for a bias smaller than the standard uncertainty is difficult. There must be at least 16 replicate measurements to make the minimum detectable bias (MDB) less than σ, the true method standard deviation, and at least 54 replicate measurements to make the MDB less than $\sigma/2$ (see MARLAP Chapter 6A). This problem seems to be inescapable for absolute bias tests based on method-blank analyses. For relative bias tests based on spiked samples, statistics can be improved if there are high- activity samples whose reference values have small uncertainties.

The mean squared error (MSE) is the sum of the squared differences between the measurements and the true values. The MSE is the sum of the variance, σ^2, and the square of the bias, b^2. The root MSE = $\sqrt{\sigma^2 + b^2}$. If the root mean squared error is kept below the required method uncertainty, the MQOs are likely to be met. If the bias is less than one third the relative method standard uncertainty, bias will only contribute 10% to the MSE.

The validation criteria in MARLAP were developed with the presumption that any known biases in the method will be corrected, and that any remaining bias will be small compared to the method uncertainty. Thus, the primary focus was placed on detecting an unacceptably high method uncertainty. Note that if the reverse is true, namely that the method uncertainty is much smaller than the bias, the method may pass the acceptance criteria while having what might be considered unacceptably high bias. In the extreme case of zero method uncertainty, a method with bias up to 30% might still pass the criteria. In this case different validation criteria may be desirable. Appendix E contains alternative method validation criteria which treat the detection of excessive bias and imprecision more equally using the concept of MSE.

APPENDIX E: AN ALTERNATIVE METHOD VALIDATION CRITERION

Introduction

The method validation process and acceptance criteria described in Section 5.4 are based on the criteria recommended in Chapter 6 of the *Multi-Agency Radiological Laboratory Analytical Protocols Manual* (MARLAP). This appendix presents alternative method performance acceptance criteria that may have greater power to detect large imprecision and bias in some situations. However, the number of test levels and replicates for the appropriate method validation level (B, C, D, E) as presented in Section 5.4.2 are still to be used.

Every measurement process involves both bias and imprecision to some degree. The MARLAP method validation criterion is predicated on the assumption that the laboratory has eliminated any substantial bias in the measurement process, so that measurement results are likely to be evenly distributed about the true value. If this assumption is not true, use of the MARLAP test alone may in some cases allow a method with a substantial bias to be accepted for use. Although MARLAP recommends that the candidate method be evaluated for bias, it does not recommend an objective criterion for determining whether a detected bias is tolerable. Furthermore, as noted in MARLAP and in Appendix D of this document, testing for bias tends to be difficult in any case, because of the number of measurements required to detect a bias that is comparable in magnitude to the standard deviation.

The assumption of this appendix is that a measurement process may be considered adequate for its intended use if a certain combination of bias and imprecision, called the "root mean squared error," does not exceed the required uncertainty. According to this view, the fact that bias is hard to quantify is less troublesome, because what one cares about most is not bias alone or imprecision alone but a combination of the two.

Definitions

Suppose $\hat{X}$ is an estimator for some parameter K. The *variance* of $\hat{X}$ denoted by $V(\hat{X})$ or $\sigma_{\hat{X}}^2$, is defined as the expected value of the square of the deviation of X from its mean.

$$\sigma_{\hat{X}}^2 = E[(\hat{X} - E(\hat{X}))^2] \tag{E1}$$

where $E(\bullet)$ denotes the expected value (mean) of the operand within the brackets or parentheses. The square root of the variance, denoted by $\sigma_{\hat{X}}$, is called the *standard deviation*. The standard deviation of an estimator is commonly used as a measure of its imprecision.

The *error* of $\hat{X}$ (as an estimator for K) is defined to be the difference between $\hat{X}$ and K.

$$Error(\hat{X}) = \hat{X} - K \qquad (E2)$$

The error $\hat{X} - K$, like $\hat{X}$ itself, is a random variable.

The *bias* of $\hat{X}$ is defined as the difference between the expected value (or mean) of $\hat{X}$ and the value of the parameter K. In symbols,

$$Bias(\hat{X}) = E(\hat{X}) - K \qquad (E3)$$

The bias of $\hat{X}$ also equals its mean error. So $Bias(\hat{X}) = E(\hat{X}) - K$

If $\hat{X}$ is an unbiased estimator (i.e., if $E(\hat{X}) - K$), then the standard deviation is a good measure of the overall quality of $\hat{X}$ as an estimator. However, in the context of laboratory analyses, the estimator $\hat{X}$ is typically the result of a measurement made using a specified method and measurement process, and in this situation, $\hat{X}$ is usually biased to some extent. It is common to evaluate a laboratory method or measurement process in terms of both the bias and imprecision (standard deviation) of the estimator $\hat{X}$. During method validation, separate limits may be set for the maximum allowable bias and for the maximum allowable standard deviation; however, since the overall quality of a measurement process is affected by both bias and imprecision, one may instead choose to specify a limit for some combination of the bias and imprecision. If this is done, then a biased but precise method may be considered to be as good as an essentially unbiased but less precise method.

Note that although neither the bias nor the standard deviation is ever known exactly, it is possible to use statistical methods to test hypotheses about their magnitudes or to determine likely bounds for their values. Note also that acknowledging the existence of bias in a measurement process does not mean that one should cease trying to find and eliminate the causes of any significant bias.

The "mean squared error" or the "root mean squared error" of an estimator is often used as a measure of the estimator's overall quality. The *mean squared error* of $\hat{X}$, as the name implies, is the expected value of the squared error of $\hat{X}$. So:

$$MSE(\hat{X}) = E[(\hat{X} - K)^2] \tag{E4}$$

Notice that the definition of $MSE(\hat{X})$ resembles that of the variance $\sigma_{\hat{X}}^2$, but with K substituted for the mean $E(\hat{X})$. It can be shown mathematically that the mean squared error of is equal to the sum of its squared bias and its variance.

$$MSE(\hat{X}) = Bias(\hat{X})^2 + \sigma_{\hat{X}}^2 \tag{E5}$$

The *root mean squared error* of $\hat{X}$ is simply the positive square root of $MSE(\hat{X})$.

$$\sqrt{MSE(\hat{X})} = \sqrt{E[(\hat{X} - K)^2]} = \sqrt{Bias(\hat{X})^2 + \sigma_{\hat{X}}^2} \tag{E6}$$

So the root mean squared error can be viewed as a mathematical combination of bias and imprecision. For an unbiased estimator, the root mean squared error is exactly equal to the standard deviation, but for a biased estimator, the root mean squared error is always larger than the standard deviation.

The approach to method validation described in this document is based on the concept of a required uncertainty at each activity level. If one interprets this required uncertainty, u, as a required bound for $\sqrt{MSE(\hat{X})}$, then an unbiased method can have a standard deviation $\sigma_{\hat{X}}$ as large as u, or a perfectly precise method can have a bias as large as u. In general, both the bias and standard deviation may be nonzero, but in principle, neither the bias nor the standard deviation is allowed to exceed the required uncertainty, u, at any level of activity.

Alternative Method Validation Criterion

The validation procedure of Section 5.4 involves making several measurements of samples spiked at known activity levels. Let L denote the number of activity levels and N the number of measurements made at each level. Then the test described in Section 5.4 compares each result X_{ij} (where i denotes the number of the activity level and j denotes the number of the measurement) to acceptance limits:

$$K_i \pm k u_i \qquad (E7)$$

where
$K_i =$ target value at the ith activity level ($1 \leq i \leq L$)
$k =$ uncertainty multiplier (from Table 3)
$u_i =$ required uncertainty at the ith analyte level

The method is judged acceptable if every result X_{ij} falls within the appropriate acceptance limits for its activity level.

As noted under Table 3, the uncertainty multiplier, k, may be calculated as follows:

$$k = z_{0.5+0.5(1-\alpha)^{1/LN}} \qquad (E8)$$

where α is the chosen significance level, or the probability of a false rejection ($\alpha = 0.05$), and for any p, z_p denotes the p-quantile of the standard normal distribution. (Note that k is rounded to two figures in Table 3.) The multiplier k also equals the square root of the $(1 - \alpha)1/LN$ -quantile of the chi-squared distribution with one degree of freedom, and for purposes of exposition, it will be convenient to use the latter interpretation here.

$$k = \sqrt{\chi^2_{(1-\alpha)^{1/LN}}(1)} \qquad (E9)$$

The required uncertainty, u_i, at each activity level equals the required method uncertainty, u_{MR}, if $K_i \leq AL$, and it equals $\varphi_{MR} K_i$ if $K_i > AL$.

$$u_i = \begin{cases} u_{MR}, & \text{if } K_i \leq AL \\ \varphi_{MR} \times K_i, & \text{if } K_i > AL \end{cases} \qquad (E10)$$

A more traditional presentation of the same statistical test would define a test statistic and a critical value for that statistic. For this test, the statistic can be defined as:

$$M = \max_{\substack{1 \le i \le L \\ 1 \le j \le N}} Z_{ij}^2 \qquad (E11)$$

where for each i and j, Z_{ij} denotes the "Z-score" for the measurement:

$$Z_{ij} = \frac{X_{ij} - K_i}{u_i} \qquad (E12)$$

The corresponding critical value for the statistic M is just the square of the uncertainty multiplier, which equals:

$$m_C = k^2 = \chi^2_{(1-\alpha)^{1/LN}}(1) \qquad (E13)$$

So the method's performance is considered acceptable if the value of M does not exceed m_C.[8]

Because the statistic M is derived from only the most extreme value of Z_{ij}, it essentially discards much of the information contained in the measurement data, resulting in reduced power for the test. A different statistic that makes better use of the same data is the following:

$$W = \max_{1 \le i \le L} \sum_{j=1}^{N} Z_{ij}^2 \qquad (E14)$$

where all the symbols on the right-hand side are as defined above for M. The critical value of the statistic W is the $(1-\alpha)^{1/L}$-quantile of the chi-squared distribution with N degrees of freedom:

$$w_c = \chi^2_{(1-\alpha)^{1/L}}(N) \qquad (E15)$$

where again $\alpha = 0.05$. This test can also be implemented by calculating a statistic W_i at each activity level:

$$W_i = \sum_{j=1}^{N} Z_{ij}^2 \qquad (E16)$$

and comparing W_i to the critical value w_C. If W_i at any activity level exceeds w_C, the method is rejected.[9]

Note that both the MARLAP test and the W test can be viewed as chi-squared tests. The MARLAP test, which is equivalent to the method validation criterion presented in Section 5.4, uses a chi-squared statistic with one degree of freedom for each of the LN measurements, while the W test pools the data for each activity level to obtain fewer statistics (L of them), each of which has more degrees of freedom (N).

Under the assumptions that the root MSE of the measurement process at each activity level does not exceed the required uncertainty, and that all the measurement results are independent, either test (MARLAP or W) will incorrectly reject a candidate method at most 5 % of the time (because $\alpha = 0.05$). The greatest false rejection rate (5 %) occurs when the bias is zero and the standard deviation at each activity level exactly equals the required uncertainty. The important differences between the two tests are the differences in their power to reject methods with larger bias or imprecision. For example, if a candidate method has negligible imprecision, either test will reject it if the measurement bias at some activity level is larger than a specified multiple of the required uncertainty. However, the associated uncertainty multiplier for the W test ($\sqrt{w_c / N}$) is generally smaller than the multiplier (k or $\sqrt{m_c}$) for the MARLAP test. Furthermore, the MARLAP test has the undesirable property that it actually loses power to detect such biases as the number of measurements (N) as each activity level increases, because the value of k increases with N. The power of the W test, on the other hand, improves, because the value of ($\sqrt{w_c / N}$) decreases with N, approaching 1 as N goes to infinity.

The Holst-Thyregod Test for Mean Squared Error

The W test described above was originally derived as a test of variance given a presumed value for the mean, but it can be employed as a test of the MSE or root MSE, as was done above. Holst and Thyregod have also derived a statistical test that explicitly tests hypotheses about the MSE of a measurement process (Holst and Thyregod, 1999). The Holst-Thyregod statistic is slightly more complicated to calculate, and tables of percentiles for the statistic are not widely available. However, the Holst-Thyregod test has an

Method Validation Guide for Qualifying Methods Used ...

advantage over the W test when the MSE is dominated by bias. In some situations where the measurement process has good precision and somewhat large bias, the power of the Holst-Thyregod test far exceeds the power of the W test, just as the power of the W test in some situations can far exceed the power of the MARLAP test. In other situations, and especially when the MSE is dominated by variance rather than bias, the W test outperforms the Holst-Thyregod test. For these reasons, this appendix recommends the W test as the best choice for most method validation experiments.

Example

Suppose the action level for a certain project is 100 pCi/L and the required method uncertainty is $u_{MR} = 10.0$ pCi/L at the action level. The relative required method uncertainty (at or above the action level) is $\varphi_{MR} = 0.10$, or 10 %. A Level D validation experiment is performed for a candidate method, with three activity levels ($L = 3$) and seven measurements at each activity level ($N = 7$). Suppose the measurement results are as shown in Table E1.

The method appears to have good precision, but it also has a relative bias of approximately −16 %, which is larger than the required relative method uncertainty (10 %).[10] If one performs the validation test described in Section 5.4, the acceptance limits for the results are as shown below.

Table E1. Method Validation Measurement Results

Measurement (j)	Target Value (Ki)		
	50 pCi/L	100 pCi/L	300 pCi/L
1	36.1	83.2	256.1
2	39	83.7	235.2
3	42.2	84.8	249
4	44.4	75.4	258.5
5	47.5	82.3	265.2
6	40.2	94.7	255.7
7	44	88.4	254.5
Average	$\bar{x} = 41.91$	$\bar{x} = 84.64$	$\bar{x} = 253.46$
Standard deviation	$s = 3.81$	$s = 5.90$	$s = 9.40$

Table E2. Acceptance Limits, MARLAP Test

Target Value K_i/(pCi/L)		Required Uncertainty u_i/(pCi/L)	Acceptance Limits $(K_i \pm k u_i)$/(pCi/L)
50		10	20 – 80
100	(AL)	10	70 – 130
300		30	210 – 390

Note:
The "uncertainty multiplier" (k) in this case equals 3.0307, which is rounded to 3.0.

Since all the measured results are within these acceptance limits, the method is judged acceptable in spite of the obvious negative bias.

If the chi-squared test described in this appendix is used instead, then the critical value for the chi-squared statistic is

$$w_c = \chi^2_{(1-\alpha)^{1/L}}(N) = \chi^2_{0.95^{1/3}}(7) = 17.07$$

and the results are shown in Table E3.

Because the statistics W_2 and W_3 both exceed the critical value $w_C = 17.1$, the method is judged to be unacceptable.

Table E3. Method Validation Results, Alternative Test (W Test)

Measurement (j)	Activity Level (i)						
	1		2		3		
	$K_i = 50$ pCi/L $u_i = 10.0$ pCi/L		$K_i = 100$ pCi/L $u_i = 10.0$ pCi/L		$K_i = 300$ pCi/L $u_i = 30.0$ pCi/L		
		$Z_{ij} = \frac{X_{ij} - K_i}{u_i}$	X_{ij}	$Z_{ij} = \frac{X_{ij} - K_i}{u_i}$	X_{ij}	$Z_{ij} = \frac{X_{ij} - K_i}{u_i}$	
1	36.1	-1.39	83.2	-1.68	256.1	-1.4633
2	39	-1.10	83.7	-1.63	235.2	-2.1600
3	42.2	-0.78	84.8	-1.52	249	-1.7000
4	44.4	-0.56	75.4	-2.46	258.5	-1.3833
5	47.5	-0.25	82.3	-1.77	265.2	-1.1600
6	40.2	-0.98	94.7	-0.53	255.7	-1.4767
7	44	-0.60	88.4	-1.16	254.5	-1.5167
	$W_i = \sum Z_{ij}^2 =$	5.45	$W_i = \sum Z_{ij}^2 =$	18.6	$W_i = \sum Z_{ij}^2 =$	17.4

Theoretical Comparison of Statistical Power

The following set of four figures graphically illustrates the power of the MARLAP test and the *W* test for the same conditions assumed in Figure D2 of Appendix D. The power of the Holst-Thyregod (H-T) test is also graphed for comparison. The scenario (as above) involves a Level D validation of a method for a project where the required method uncertainty is 10 pCi/L at an action level of 100 pCi/L. Each of the following figures assumes a different value for the ratio of the relative standard deviation (RSD) to the required uncertainty at each activity level. In Figure E1a, the ratio is 0.5, so that the RSD at the action level equals 5 %. The ratios for Figures E1b, E1c, and E1d are 0.75, 1, and 1.25, respectively. In each graph, the horizontal axis represents possible values for the relative bias of the method ranging from 0 to 20 %. The vertical axis, labeled *P*, represents the probability that a method with the given relative standard deviation and relative bias will be rejected.

In every case, the *W* test outperforms the MARLAP test, although the differences are most noticeable when the precision of the method is good but the bias is large. Also note that the power of the HolstThyregod test exceeds that of the *W* test in Figures E1a and E1b but not in E1c and E1d.

Reference

[1] Holst, Erik & Poul Thyregod (1999). "A statistical test for the mean squared error," *Journal of Statistical Computation and Simulation*, 63:4, 321–347

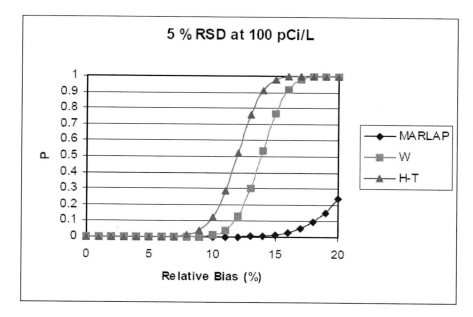

Figure E1a.

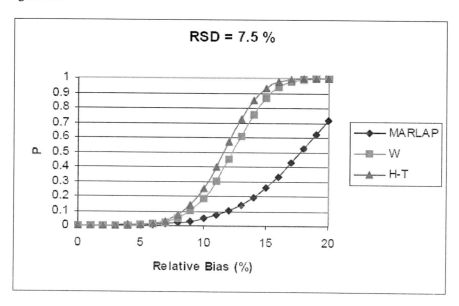

Figure E1b.

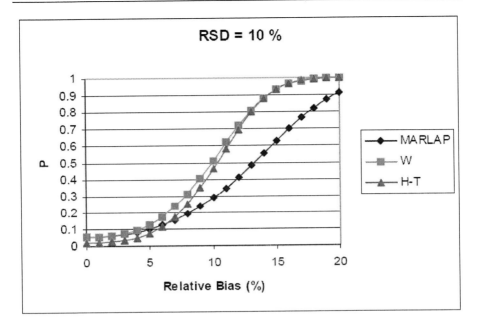

Figure E1c.

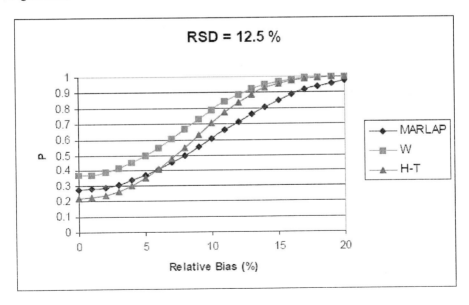

Figure E1d.

APPENDIX F: GLOSSARY

accuracy: The closeness of a measured result to the true value of the quantity being measured. Various recognized authorities have given the word "accuracy" different technical definitions, expressed in terms of bias and imprecision. Following MARLAP, this document avoids all of these technical definitions and uses the term "accuracy" in its common, ordinary sense.

aerosol: A suspension of fine solid or liquid particles within a gaseous matrix (usually air).

aliquant: A representative portion of a homogeneous *sample* removed for the purpose of analysis or other chemical treatment. The quantity removed is not an evenly divisible part of the whole sample. An aliquot, by contrast, is an evenly divisible part of the whole.

analyte: For this document, an analyte is a specific radionuclide or a category of radionuclides that comprise gross alpha or beta analyses. An analyte may be on the list of radionuclides of interest or a radionuclide of concern for a project. See *target analyte*.

analyte concentration range: (1) Method validation definition – the radionuclide concentration range corresponding to three test levels (lower, mid and upper) that are used during method validation. The mid level concentration corresponds to the action level. (2) MQO definition – the expected concentration range (minimum to maximum) of an *analyte* expected to be present in a *sample* for a given project. While most analytical protocols are applicable over a fairly large range of concentration for the *radionuclide of interest*, performance over a required concentration range can serve as a measurement quality objective for the protocol selection process, and some analytical protocols may be eliminated if they cannot accommodate the expected range of concentration.

analytical action level (**AAL**): The value of a quantity that will cause the decision maker to choose one of the alternative actions. The action level may be a derived concentration level (such as the *derived water concentration* in this document), background level, release criteria, regulatory decision limit, etc. The AAL is often associated with the type of media, *target analyte*, and concentration limit. Some AALs are expressed in terms of a derived radionuclide concentration corresponding to a dose or risk, such as a protective action guide. MARLAP uses the term "action level."

analytical decision level (ADL): The minimum measured value for the radionuclide concentration in a sample that indicates the amount of radionuclide present is equal to or greater than the *analytical action level* at a specified *Type II error* rate. (Assumes that *method uncertainty* requirements have been met.) Any measurement result equal to or greater than the applicable ADL is considered to have exceeded the corresponding *analytical action level*. MARLAP uses the term *"critical level."*

analytical protocol specification (APS): The output of a *directed planning process* that contains the project's analytical data needs and requirements in an organized, concise form. The level of specificity in the APSs should be limited to those requirements that are considered essential to meeting the project's *analytical data requirements* to allow the laboratory the flexibility of selecting the protocols or methods that meet the analytical requirements.

background (instrument): Radiation detected by an instrument when no *source* is present. The background radiation that is detected may come from radionuclides in the materials of construction of the detector, its housing, its electronics, and the building, as well as the environment and natural radiation.

background level: A term that usually refers to the presence of radioactivity or radiation in the environment. From an analytical perspective, the presence of background radioactivity in samples needs to be considered when clarifying the radioanalytical aspects of the decision or study question. Many radionuclides are present in measurable quantities in the environment.

bias (of a measurement process): A persistent deviation of the mean measured result from the true or accepted reference value of the quantity being measured, which does not vary if a measurement is repeated.

blank (analytical or method): A *sample* that is assumed to be essentially free of the *target analyte* (the "unknown"), which is carried through the radiochemical preparation, analysis, mounting, and measurement process in the same manner as a routine sample of a given matrix.

calibration: The set of operations that establish, under specified conditions, the relationship between values indicated by a measuring instrument or measuring system, or values represented by a material measure, and the corresponding known value of a parameter of interest.

calibration source: A prepared *source*, made from a *certified reference material* (standard), that is used for calibrating instruments.

carrier: (1) A stable isotopic form of a tracer element or nonisotopic material added to effectively increase the quantity of a tracer element during radiochemical procedures, ensuring conventional behavior of the element in solution. (2) A substance in appreciable amount that, when associated with a tracer of a specified substance, will carry the tracer with it through a chemical or physical process, or prevent the tracer from undergoing non-specific processes due to its low concentration (IUPAC, 1995). A stable isotope of a *radionuclide* (usually the *analyte*) added to increase the total amount of that element so that a measurable mass of the element is present.

chain of custody: Procedures that provide the means to trace the possession and handling of a sample from collection to data reporting.

check source: A material used to validate the operability of a radiation measurement device, sometimes used for instrument quality control. See *source, radioactive*.

combined standard uncertainty: Standard uncertainty of an *output estimate* calculated by combining the standard uncertainties of the *input estimates*. The *combined standard uncertainty* of y is denoted by $u_c(y)$. See *uncertainty (of measurement)*.

critical level: Termed *analytical decision level* in this document in the context of evaluating sample results relative to an *analytical action level*. In the context of analyte detection, *critical level* means the minimum measured value (e.g., of the instrument signal or the radionuclide concentration) that indicates a positive (nonzero) amount of a radionuclide is present in the material within a specified probable error. The critical level is sometimes called the *critical value* or *decision level*.

critical net concentration: Similar in concept as the "critical level."

data quality objective (DQO): Qualitative and quantitative statements that clarify the study objectives, define the most appropriate type of data to collect, determine the most appropriate conditions from which to collect the data, and specify tolerable limits on decision error rates. Because DQOs will be used to establish the quality and quantity of data needed to support decisions, they should encompass the total *uncertainty* resulting from all data collection activities, including analytical and sampling activities.

default AAL test level: Radionuclide test concentration for a given general matrix category to be used in the method validation process in the absences of PAG or risk-based AALs.

derived air concentration (DAC): The concentration of a radionuclide, in pCi/m^3, that would result in exposure to a specified dose level. Generally refers to a *protective action guide* or other specified dose- or risk-based factor related to an *analytical action level*. In this document, for example, the "500-mrem DAC for ^{239}Pu" is the concentration of ^{239}Pu, in pCi/m^3, that would result in an exposure of 500 mrem and would refer to the 500-mrem PAG. The DAC is radionuclide-specific.

derived radionuclide concentration (DRC): General application term used in discussions involving both of the terms DAC and DWC.

derived water concentration (DWC): The concentration of a radionuclide, in pCi/L, that would result in exposure to a specified dose level. Generally refers to a *protective action guide* or other specified dose- or risk-based factor related to an analytical action level.

detection capability: The capability of a *measurement process* to distinguish small amounts of *analyte* from zero.

detection limit: The smallest value of the amount or concentration of *analyte* that ensures a specified high probability of detection. Also called "*minimum detectable value.*"

discrimination limit (DL): The DL is the point where it is important to be able to distinguish expected signal from the *analytical action level*. The boundaries of the *gray region*.

dose equivalent: Quantity that expresses all radiations on a common scale for calculating the effective absorbed dose. This quantity is the product of absorbed dose (*grays* (Gy) or rads) multiplied by a quality factor and any other modifying factors (MARS SIM, 2000). The quality factor adjusts the absorbed dose because not all types of ionizing radiation create the same effect on human tissue. For example, a dose equivalent of one *sievert* (Sv) requires 1 Gy of beta or gamma radiation, but only 0.05 Gy of alpha radiation or 0.1 Gy of neutron radiation. Because the sievert is a large unit, radiation doses often are expressed in millisieverts (mSv). See *total effective dose equivalent* and *roentgen*.

gray (Gy): The International System of Units (SI) unit for absorbed radiation dose. One gray is 1 joule of energy absorbed per kilogram of matter, equal to 100 *rad*. See *sievert*.

gray region: The range of possible values in which the consequences of decision errors are relatively minor. Specifying a gray region is necessary because variability in the analyte in a population and imprecision in the measurement system combine to produce variability in the data such that the decision may be "too close to call" when the true value is very near the *analytical action level*. The *gray region* establishes the minimum distance from the *analytical action level* where it is most important to control *Type II decision errors*.

hypothesis testing: The use of statistical procedures to decide whether a null hypothesis should be rejected in favor of an *alternative hypothesis* or not rejected.

incident response method validation: Project method validation for incident response applications. See *project method validation* and *method validation*.

interferences: The presence of other chemicals or *radionuclides* in a *sample* that hinder the ability to analyze for the *radionuclide of interest*.

MARLAP Process: A performance-based approach that develops *Analytical Protocol Specifications*, and uses these requirements as criteria for the analytical protocol selection, development, and evaluation processes, and as criteria for the evaluation of the resulting laboratory data. This process, which spans the three phases of the *data life cycle* for a project, is the basis for achieving MARLAP's basic goal of ensuring that radioanalytical data will meet a project's or program's data requirements or needs.

measurand: "Particular quantity subject to measurement" (ISO, 1993a).

measurement quality objective (MQO): The analytical data requirements of the *data quality objectives*, which are project- or program-specific and can be quantitative or qualitative. These analytical data requirements serve as measurement performance criteria or objectives of the analytical process. MARLAP refers to these performance objectives as MQOs. Examples of quantitative MQOs include statements of required analyte detectability and the *uncertainty* of the analytical protocol at a specified radionuclide concentration, such as the action level. Examples of qualitative MQOs include statements of the required specificity of the analytical protocol (e.g., the ability to analyze for the radionuclide of interest [or *target analyte*] given the presence of interferences).

measurement uncertainty: See *uncertainty*.

method blank: A *sample* assumed to be essentially *target analyte*-free that is carried through the radiochemical preparation, analysis, mounting and measurement process in the same manner as a routine sample of a given matrix.

method performance characteristics: The characteristics of a specific *analytical method* such as *method uncertainty*, *method range*, *method specificity*, and *method ruggedness*. MARLAP recommends developing *measurement quality objectives* for select *method performance characteristics*, particularly for the *uncertainty (of measurement)* at a specified concentration (typically the *action level*).

method ruggedness: The relative stability of method performance for small variations in method parameter values.

***method specificity*:** The ability of the method to measure the *analyte* of concern in the presence of interferences.

***method uncertainty*:** Refers to the predicted *uncertainty* of the result that would be measured if the method were applied to a hypothetical laboratory *sample* with a specified analyte concentration. Although individual measurement uncertainties will vary from one measured result to another, the *required method uncertainty* is a target value for the individual measurement uncertainties and is an estimate of uncertainty before the sample is actually measured.

***method validation (MV)*:** The demonstration that the method selected for the analysis of a particular analyte in a given matrix is capable of providing analytical results to meet the project's *measurement quality objectives* and any other requirements in the analytical protocol specifications.

***minimum detectable concentration (MDC)*:** An estimate of the smallest true value of the analyte concentration that gives a specified high probability of detection.

***nuclide-specific analysis*:** Radiochemical analysis performed to isolate and measure a specific radionuclide.

***null hypothesis (H_0)*:** One of two mutually exclusive statements tested in a statistical *hypothesis test* (compare with alternative hypothesis). The null hypothesis is presumed to be true unless the test provides sufficient evidence to the contrary, in which case the *null hypothesis* is rejected and the alternative hypothesis (H_1) is accepted.

performance evaluation (PE) program: A laboratory's participation in an internal or external program of analyzing proficiency-testing samples appropriate for the analytes and matrices under consideration (i.e., PE program traceable to a national standards body, such as NIST). Referencematerial samples used to evaluate the performance of the laboratory may be called performanceevaluation, performance or proficiency-testing samples or materials. See *proficiency test samples*.

precision: The closeness of agreement between independent test results obtained by applying the experimental procedure under stipulated conditions.

Precision may be expressed as the standard deviation. Conversely, imprecision is the variation of the results in a set of replicate measurements.

proficiency test (PT) samples: Samples having a known radionuclide concentration used in a PE program or internally at the laboratory for method validation and for the measurement of bias.

project method validation: The demonstrated method applicability for a particular project.

protective action guide (PAG): The radiation dose to individuals in the general population that warrants protective action following a radiological event. In this document, PAGs limit the projected radiation doses for different exposure periods: not to exceed 2-rem *total effective dose equivalent (TEDE)* during the first year, 500-mrem TEDE during the second year, or 5 rem over the next 50 years (including the first and second years of the incident). See *derived water concentration and analytical action level.*

quality control (QC): The overall system of technical activities that measures the attributes and performance of a process, item, or service against defined standards to verify that they meet the stated requirements established by the project; operational techniques and activities that are used to fulfill requirements for quality. This system of activities and checks is used to ensure that measurement systems are maintained within prescribed limits, providing protection against out-of- control conditions and ensuring that the results are of acceptable quality.

radiochemical analysis: The analysis of a sample matrix for its radionuclide content, both qualitatively and quantitatively. radionuclide: A nuclide that is radioactive (capable of undergoing radioactive decay).

relative required method uncertainty (φ_{MR}): The required method uncertainty divided by the analytical action level. The relative required method uncertainty is applied to radionuclide concentrations above the analytical action level. A key measurement quality objective.

rem: The common unit for the effective or equivalent dose of radiation received by a living organism, equal to the actual dose (in rads) multiplied by a factor representing the danger of the radiation. Rem is an abbreviation for "roentgen equivalent man," meaning that it measures the biological effects of ionizing radiation in humans. One rem is equal to 0.01 Sv. See *sievert*.

replicates: Two or more *aliquants* of a homogeneous sample whose independent measurements are used to determine the *precision* of laboratory preparation and analytical procedures.

required method uncertainty (u_{MR}): *Method uncertainty* at a specified concentration. A *key measurement quality objective*. See *relative required method uncertainty*.

required minimum detectable concentration (RMDC): An upper limit for the minimum detectable concentration required by some projects.

sample: (1) A portion of material selected from a larger quantity of material. (2) A set of individual samples or measurements drawn from a population whose properties are studied to gain information about the entire population.

screening method: An economical gross measurement (alpha, beta, gamma) used in a tiered approach to method selection that can be applied to analyte concentrations below an *analyte* level in the *analytical protocol specifications* or below a fraction of the specified *action level*.

sievert (Sv): The SI unit for the effective dose of radiation received by a living organism. It is the actual dose received (grays in SI or rads in traditional units) times a factor that is larger for more dangerous forms of radiation. One Sv is 100 *rem*. Radiation doses are often measured in mSv. An effective dose of 1 Sv requires 1 gray of beta or gamma radiation, but only 0.05 Gy of alpha radiation or 0.1 Gy of neutron radiation.

swipe: A filter pad used to determine the level of general radioactive contamination when it is wiped over a specific area, about 100 cm^2 in area. Also called "smear" or "wipe."

target analyte: A radionuclide on the list of radionuclides of interest or a radionuclide of concern for a project. For incident response applications, typical radionuclides of interest are provided in Appendix A.

total effective dose equivalent: The sum of the effective dose equivalent (for external exposure) and the committed effective dose equivalent (for internal exposure), expressed in units of Sv or rem.

Type I decision error: In a hypothesis test, the error made by rejecting the null hypothesis when it is true. A Type I decision error is sometimes called a "false rejection" or a "false positive."

Type II decision error: In a hypothesis test, the error made by failing to reject the null hypothesis when it is false. A Type II decision error is sometimes called a "false acceptance" or a "false negative."

uncertainty: A parameter, usually associated with the result of a measurement, that characterizes the dispersion of the values that could reasonably be attributed to the *measurand*.

End Notes

[1] Throughout this guide, the term "Incident Commander" (or "IC") includes his or her designee.

[2] Here, β means the probability of a Type II decision error.

[3] The term "decontamination factor" is defined as the amount of interferent in the sample before chemical separation divided by the measured amount in the sample after chemical separation.

[4] The IC may develop and require other AALs and required method uncertainties. If so, the IC should verify whether the laboratory can meet the new method uncertainty requirements for the updated AALs. Calculating the test levels for method validation should be consistent with Table 2.

[5] Notice that the sum under the radical in equation 4 is divided by the number of degrees of freedom, $N - 1$, not the number of results, N. When calculated in this manner, s_X^2 is an unbiased estimator for the variance of the results. If the true mean of the results, μ_X, were known, a better estimate of the variance would be, $\frac{1}{N}\sum_{i=1}^{N}(X_i - \mu_x)^2$, but because 1 the N 1 Xi i x N(-) = Σ μ mean is estimated from the data, the number of degrees of freedom is reduced by 1. Notice also that the expression in the denominator of the right-hand side of Equation 5 gives the experimental standard deviation of the mean, more commonly known as the "standard error of the mean." The division by $\sqrt{N}$ in this case accounts for the effect of averaging N independent results.

[6] During the preparation of the proficiency test samples for a test level, the spread in activity deposited on the samples of the test level should be controlled so that the coefficient of variation of the test-sample activities does not exceed 3%.

[7] EPA 2008. *Radiological Laboratory Sample Analysis Guide for Incidents of National Significance – Radionuclides in Water*, Table 9A.

[8] The test could also be based on a statistic equal to the maximum of the absolute values $|Z_{ij}|$, using k as a critical value.

[9] The expected value of W_i equals N times the ratio of the mean squared error (MSE) to the square of the required uncertainty (u_i^2) at this activity level.

[10] Presumably the laboratory was unaware of this bias; otherwise, it would have corrected it.

In: National Radiation Incidents: Laboratory ... ISBN: 978-1-61324-666-5
Editor: Martin C. Sheckley © 2011 Nova Science Publishers, Inc.

Chapter 2

RADIOLOGICAL LABORATORY SAMPLE SCREENING ANALYSIS GUIDE FOR INCIDENTS OF NATIONAL SIGNIFICANCE[*]

Environmental Protection Agency

ACKNOWLEDGMENTS

This manual was developed by the National Air and Radiation Environmental Laboratory (NAREL) of EPA's Office of Radiation and Indoor Air (ORIA). Dr. John Griggs was the project lead for this document. Several individuals provided valuable support and input to this document throughout its development. Special acknowledgment and appreciation are extended to Dr. Keith McCroan, ORIA/NAREL; Mr. Daniel Mackney for instrumental sample analysis support, ORIA/NAREL; Ms. Schatzi Fitz-James, Office of Emergency Management, Homeland Security Laboratory Response Center; and Mr. David Garman, ORIA/NAREL. We also wish to acknowledge the external peer reviews conducted by Lindley J. Davis and Carolyn Wong, whose thoughtful comments contributed greatly to the understanding and quality of the report. Numerous other individuals, both inside and outside of EPA, provided peer review of this document, and their suggestions contributed greatly to the quality and consistency of the final document. Technical support

---

[*] This is an edited, reformatted and augmented version of United States Environmental Protection Agency Report EPA 402-R-09-008, dated June 2009.

was provided by Dr. N. Jay Bassin, Dr. Anna Berne, Mr. David Burns, Dr. Carl V. Gogolak, Dr. Robert Litman, Dr. David McCurdy, and Mr. Robert Shannon of Environmental Management Support, Inc.

ACRONYMS, ABBREVIATIONS, UNITS AND SYMBOLS

(Excluding chemical symbols and formulas)

α	alpha particle
AAL	analytical action level
ADL	analytical decision level
AL	action level
β	beta particle
Bq	becquerel (1 dps)
CERCLA	Comprehensive Environmental Response, Compensation, and Liability Act of 1980 ("Superfund")
cfm	cubic feet per minute
CFR	*Code of Federal Regulations*
cm	centimeter
cpm	counts per minute
d	day
DAC	derived air concentration
DL	discrimination limit
DOE	U.S. Department of Energy
DP	decay product(s)
dpm	disintegration per minute
dps	disintegration per second
DQO	data quality objective
DRP	discrete radioactive particle
e^-	electron
$E_{\beta max}$	maximum energy of the beta-particle emission
EDD	electronic data deliverable
EPA	U.S. Environmental Protection Agency
ERLN	Environmental Response Laboratory Network
FOM	figure of merit
γ	gamma ray
g	gram
Ge	germanium [semiconductor]

GM	Geiger-Muller detector
GP	gas proportional
GPC	gas proportional counting [counter]
GS	gamma spectrometry
Gy	gray
h	hour
H_0	null hypothesis
H_1	alternate hypothesis
HPGe	high-purity germanium detector
IC	Incident Commander [or designee]
ICLN	Integrated Consortium of Laboratory Networks
IND	improvised nuclear device
INS	incident of national significance
keV	thousand electron volts
L	liter
LBGR	lower bound of the gray region
LCS	laboratory control sample
LEPD	low-energy photon detector
LS	liquid scintillation
LSC	liquid scintillation counter
MARLAP	*Multi-Agency Radiological Laboratory Analytical Protocols Manual*
MARS SIM	*Multi-Agency Radiation Survey and Site Investigation Manual*
MCL	maximum contaminant level
MDC	minimum detectable concentration
MeV	million electron volts
mg	milligram (10^{-3} g)
min	minute
mL	milliliter (10^{-3} L)
MQO	measurement quality objective
mR	milliroengten (10^{-3} R)
mrem	millirem (10^{-3} rem)
μg	microgram (10^{-6} g)
NaI(Tl)	thallium-activated sodium iodide detector
NORM	naturally occurring radioactive materials
φ_{MR}	relative method uncertainty
PAG	Protective Action Guide
pCi	picocurie (10^{-12} Ci)

QA	quality assurance	
QC	quality control	
rad	radiation absorbed dose	
RDD	radiological dispersal device (i.e., "dirty bomb")	
RDL	required detection limit	
REGe	reverse electrode germanium detector	
rem	roentgen equivalent man	
RFA	responsible federal agency	
s	second	
SI	International System of Units	
SOP	standard operating procedure	
STS	sample test source	
Sv	sievert	
$t\ \frac{1}{2}$	half-life	
TAT	turnaround time	
TEDA	triethylenediamine	
TEDE	total effective radiation dose equivalent	
UBGR	upper bound of the gray region	
u_{MR}	required method uncertainty	
y	year	

RADIOMETRIC AND GENERAL UNIT CONVERSIONS

To Convert	To	Multiply by	To Convert	To	Multiply by
years (y)	seconds (s) minutes (min) hours (h) days (d)	3.16×10^7 5.26×10^5 8.77×10^3 3.65×10^2	s min h d	y	3.17×10^{-8} 1.90×10^{-6} 1.14×10^{-4} 2.74×10^{-3}
disintegrations per second (dps)	Becquerels (Bq)	1	Bq	dps	1
Bq Bq/kg Bq/m³ Bq/m³	picocuries (pCi) pCi/g pCi/L Bq/L	27.0 2.70×10^{-2} 2.70×10^{-2} 10^{-3}	pCi pCi/g pCi/L Bq/L	Bq Bq/kg Bq/m³ Bq/m³	3.70×10^{-2} 37.0 37.0 10^3
microcuries per milliliter (µCi/mL)	pCi/L	10^9	pCi/L	µCi/mL	10^{-9}
disintegrations per minute (dpm)	µCi pCi	4.50×10^{-7} 4.50×10^{-1}	pCi	dpm	2.22

Radiological Laboratory Sample Screening Analysis Guide ...

To Convert	To	Multiply by	To Convert	To	Multiply by
cubic feet (ft^3)	cubic meters (m^3)	2.83×10^{-2}	m^3	ft^3	35.3
gallons (gal)	liters (L)	3.78	liters	gallons	0.264
gray (Gy)	rad	10^2	rad	Gy	10^{-2}
roentgen equivalent man (rem)	sievert (Sv)	10^{-2}	Sv	rem	10^2

Note: Traditional units are used throughout this document instead of SI units. Protective Action Guides (PAGs) and their derived concentrations appear in official documents in the traditional units and are in common usage. Conversion to SI units will be aided by the unit conversions in this table.

I. INTRODUCTION

A response to a release of radioactivity to the environment likely will Most laboratories do not routinely screen samples under conditions found during an emergency response situation, such as from a radiological or nuclear incident of national significance (INS). Many of these samples are higher in activity and need to be accurately surveyed and prioritized for analysis based on direction from the Incident Commander (IC).[1] This document describes methods that may be applied by personnel at a radioanalytical laboratory for screening of samples for radioactivity. The specific techniques described in this guide may be used to assess the gross α, β, or γ activity in samples that may have been contaminated as the result of a radiological or nuclear event, such as a radiological dispersion device (RDD), improvised nuclear device (IND), or an intentional release of radioactive materials into the atmosphere or a body of water or aquifer, or to terrestrial areas via mechanical or other methods. In the event of a major incident that releases radioactive materials to the environment, EPA will turn to selected radio- analytical laboratories to support its response and recovery activities. In order to expedite sample analyses and data feedback, the laboratories will need guidance occur in three phases that are generally defined in this document as: "early" (onset of the event to about day 4), "intermediate" (about day 4 to about day 30), and "recovery" (beyond about day 30). Each phase of an incident response will require different and distinct radioanalytical resources to address the different consequences, management, priorities, and requirements of a phase. Some of the more important radioanalytical laboratory issues germane to an incident response consist of radionuclide identification and quantification capability, sample load capacity, sample processing turnaround time, quality of analytical data, and data transfer capability. This guide emphasizes the laboratory screening of samples from the end of the early phase, through the intermediate phase, and into the recovery phase (but does not address the screening by initial responders).

Although not the focus of this document, during the early phase, analytical priorities need to address the protection of the public and field personnel due to potentially high levels of radioactivity and the need to provide for *qualitative* identification of radionuclides. During this phase, the Protective Action Guides (PAGs) for radiological emergencies require evacuation of a population if the projected short-term total effective radiation dose equivalent[2] (TEDE) exceeds 1 rem.[3] The nominal trigger for sheltering is 1 rem over four days (projected avoided inhalation dose). The radioanalytical resource requirements (field or fixed laboratory) for this early phase may vary significantly depending on the time frame, source-term nuclide, and the extent of the contamination.

During the intermediate phase, the radionuclides and matrices of concern are known *qualitatively*, and the *quantitative* levels suitable for making decisions based on action levels need to be rapidly determined. For the intermediate phase, PAGs have been established to limit the projected radiation doses for different exposure periods, not to exceed 2-rem TEDE over the first year, 500-mrem TEDE during the second or any subsequent year, or 5 rem over the next 50 years (including the first and second years of the incident). In addition, radionuclide concentration limits for food and water as regulated by the Food and Drug Administration and EPA would be applicable.

The final, or "recovery," phase occurs as part of a radiological incident site-remediation effort. During this phase, when site atmospheric characterization and remediation cleanup effectiveness are determined, there is potential for more extensive radiochemical analyses at the lowest radionuclide concentrations.

The analytical resources needed during any phase of a radiological event will depend on the radionuclide analytical action level (AAL)[4] developed for the various media that may affect human exposure. The radionuclide AALs, which are derived radionuclide concentrations for the different media types based on the PAGs or risk values, may change depending upon the phase of the event.

The time period of an incident where this document will find its greatest utility is early in the intermediate phase through the end of the recovery phase. Laboratories performing analyses must focus on optimizing sample analyses so that the initial qualitative aspects and concentrations related to the appropriate AALs can be determined quickly (i.e., rapid turnaround of sample results). Radioanalytical screening by laboratories during these phases will include methods for all three radioactive emissions. During the recovery phase, however, the screening techniques used for samples will be more focused because the radionuclides from the event are likely to have already been identified and chemically characterized.

During all phases of an incident response, radioanalytical resources are needed for the gross radiation screening of samples for prioritization of sample processing or for information related to the general level of contamination, identification of the radionuclide source term, and quantification of the radionuclides in a variety of sample media. This document has been developed to provide guidance during an incident on techniques to enhance the ability to differentiate radioactivity in samples near action levels and optimize the calibration of the screening equipment used for gross sample activity measurement. Using these techniques should help laboratories to prioritize samples in a timely fashion based on the request of the IC.

The process of screening samples using a survey instrument can be described in two stages. The first stage deals with the receipt of the bulk sample shipment and assessment of the radiation dose rate (mrem/h) or gross activity (cpm) from the shipment and the individual samples, *prior to opening any samples*. The main purpose in this stage is to identify any immediate radiological hazard to the receipt personnel and sample analysts. This screening measurement typically is made using an instrument that does not discriminate particle energies or assess total dose rate from the sample. For example, an instrument like a Geiger-Mueller (GM) detector is sensitive to all gamma and beta particles with enough energy to pass through the container walls without identifying which is which. At this time, no assessment of alpha particle or low-energy beta particle contamination can be made. The measurement should not take more than 5 to 10 seconds to complete per sample. Important aspects of the outcome of this measurement are that the samples can be appropriately shielded and labeled for both radiation protection and prioritization purposes, and that the sample mass and integrity remain unchanged (this is a non-destructive, non-invasive test).

The second stage of screening is more substantive in that it examines the total radionuclide activity for a particular type of particle emitted from the radionuclides contained within the sample. Ideally, if ^{90}Sr, ^{14}C, and ^{99}Tc were all contained in a sample, the instrument used for screening would measure the total contribution as *the sum of the three*, even if it could not identify them individually. Unfortunately, the instruments used for *screening* are not ideal:

detector response tends to be proportional to the characteristic energy of the radiation emitted by a radionuclide and the detection is also impacted by sample self-shielding. It is very important to ensure that a screening test will provide a conservative estimate of the total activity of the radionuclides present to ensure that the screen does not underestimate the total amount of a radionuclide present. If the identity of the radionuclides is known, a different response factor should be applied when measuring the medium-to-high energy beta from ^{90}Sr/^{90}Y than for the lower energy ^{14}C in samples where mass attenuation may be significant.

Using gas proportional counting (GPC) or liquid scintillation counting (LSC) to perform the screening process has several important consequences. First, when the sample container itself is opened, the potential exists for contaminating both the sample and the laboratory. Second, a portion of the sample may be sacrificed for the screening process, which may require judicious sub-sampling. Third, chain-of-custody must be established for open sample containers and aliquanting prior to actual analysis. This will prevent questions later on regarding the sample integrity.

This document provides technical information and recommendations for a laboratory faced with screening samples received following a radiological INS. Screening samples deals with the detector responses to radiation and the effects of different forms of radiation on different detector types. Three appendices provide detailed scenarios that use the information in the technical section of the document. These scenarios illustrate when to change calibration and screening techniques based upon what is known about the sample's radioactive contaminants and the instrument detection efficiency. The methods demonstrated by these scenarios are:

- Preparation of laboratory screening equipment for an INS event;
- Receipt of samples from an INS event with known radionuclides for which the laboratory screening instruments are calibrated; and
- Receipt of samples from an INS event with known radionuclides, but the laboratory screening equipment must use a detection correction factor because the instruments were not calibrated with radionuclides present in the event samples.

Facility personnel should use these examples as guidance to prepare the screening instruments that are commonly used in their laboratories to analyze gross activity in samples from an INS.

A. Purpose and Objectives

This document describes how to develop laboratory methods to perform gross radioactivity analysis for samples resulting from an INS. It discusses technical issues associated with screening measurements, provides the suggested methodologies to determine correction factors for these instruments, offers a consistent methodology for measuring sample gross activity concentrations, and provides guidance on the calibration of screening equipment commonly used by laboratories.

Although the list of potential threat radionuclides is relatively short, instrument responses to the different particle energies may vary significantly depending upon the type of screening instrument used. It is important to be able to use screening instrumentation to support the overall laboratory process of sample prioritization and analysis that will support decisions to protect the health and safety of the public.

This document provides guidance for a user to select appropriate methods for screening at different points in the analytical process. The critical points in the process are sample receipt, sample prioritization, and rapid feedback to the IC on samples exceeding action levels.

The specific objectives for response personnel to accomplish in preparing their laboratories for such an event include:

- Performance of method validation for each instrument/sample geometry combination used in screening;
- Identification of consistent methods of screening for various media;
- Screening instrument configurations that streamline the screening process;
- Screening measurements that will aid in prioritizing samples for analyses; and
- Methods for calibration of screening equipment that will have the widest applicability to those radionuclides most likely to result from an INS.

B. Scope of DQOs/MQOs for the Screening Process

The use of screening instrumentation to prioritize samples based on the amount of activity in an individual sample should be consistent for *all* laboratories responding to an INS. This should allow the processing of samples and return of results to the IC based on the measurement quality objectives (MQOs) of the event in the timeliest manner. During the early phase of an event when the identity and extent of radioactive contamination are unknown, the screening instrumentation should be calibrated with radionuclides that are routinely used for gross screening calibrations, but in a geometry that should support the best discrimination of activity levels. As the event progresses and the specific radionuclides are identified, either the calibration may be changed to reflect the known radionuclides or an interpolated correction factor for instrument response due to other radionuclides based on energy should be used.

Other guides in this series[5] identify Protective Action Guides (PAGs) as associated concentrations and AALs that are critical measurement limits. The screening instruments used in the laboratory to support the rapid and organized evaluation of sample priority should be calibrated for gross activity measurements at these critical measurement limits in order to achieve the established MQOs stated in the other guides.

Samples that have the potential for considerations in a criminal investigation must be handled separately, and the laboratory should receive information from the Incident Commander on how to process these samples.

C. Measurement Quality Objectives: Relationship of Derived Concentrations, AAL, ADL, Risk Levels, and u_{MR} to Dose

MQOs External to the Laboratory

Gross activity screening of samples is the first step to assessing whether or not a particular sample exceeds a PAG's derived radionuclide concentration for the matrix that is being assessed for radioactive materials. PAGs establish radiation dose limits applicable to different phases of an incident response. The PAG (expressed as a numerical dose level) indicates a level of exposure at which protective action should be taken to prevent, reduce, or limit a person's radiation dose during a radiological incident. The measurements that are made with screening instruments in the radioanalytical laboratory should be correlated to the PAGs expressed as concentrations (or other AALs) for each matrix defined by the incident.

A *derived concentration* of a radionuclide that corresponds to a PAG or risk-based dose in a specific matrix can be calculated and used to facilitate the application of these radioanalytical action levels in the laboratory for decision-making purposes. For example, the derived air concentration (DAC, in units of pCi/m^3) of an individual radionuclide in air corresponds to a radiation dose (PAG) to a specific population. For each matrix that undergoes screening, there should be a derived radionuclide concentration that may be *directed* by regulation or *selected* based on the specific incident.

Screening instruments, when configured properly, can be used to conservatively determine if a sample has or has not exceeded an AAL. However, when the total gross screening activity exceeds an AAL, it may not be possible to determine if the AAL for an individual radionuclide is actually exceeded until radionuclide-specific methods are performed. In cases where it is not possible to determine if an individual radionuclide AAL has been exceeded, screening provides the laboratory with the information to prioritize samples that need to be analyzed first. The priority for sample analyses will be decided based upon the incident phase and the specific needs of the IC. For example, the order of analysis could be based on highest activity first, lowest activity first, gamma response first, or any such logical priority.

MQOs Internal to the Laboratory

The laboratory also needs the screening equipment to correlate to MQOs established in the laboratory and thus facilitate sample processing. The screening MQO will likely change as the event progresses and the known concentration of the radionuclides involved becomes more certain and their concentration diminishes due to radioactive decay, dilution, or dispersion. Using *Radiological Laboratory Sample Analysis Guide for Incidents of National Significance–Radionuclides in Air* as an example, four different levels are assessed over the course of an event: 2 rem, 500 mrem, 10^{-4} risk, and 10^{-6} risk. As the event progresses towards samples being analyzed at the level of 10^{-6} risk, the method detection capability may need to improve in order to continually and efficiently prioritize samples. The feedback to the IC will be slowed down because the decreased sample activities will result in longer screening times for samples and longer count times for samples following analytical separations.

The changing MQOs will have a "domino effect" on laboratory QC analyses, such as spikes, duplicates, laboratory control samples (LCSs), and blank samples, processed in a batch. The activity levels for spikes and LCSs may become lower as the event progresses, and the acceptance criteria for the QC samples also may change. Changes to the required measurement uncertainties for these QC samples will require longer counting times and also may slow down reporting to the IC.

The required method uncertainty (u_{MR}) may have default values for each radionuclide and matrix (other guides in this series identify these default values; see references in Appendix IV) or may have incident-driven values. In either case, the laboratory should be prepared to adjust these values when required by the incident MQOs for both the screening instruments and the radionuclide specific methods. The value of u_{MR} and the acceptable error rates for Type I and Type II errors are used to determine the analytical decision level (ADL). The ADL is a value that is less than the AAL. When the ADL is exceeded, it is concluded that the AAL has also been exceeded, guarding against a decision error that would allow a sample exceeding the AAL to go undetected. The ADL concept is also used for both screening instruments and laboratory-specific methods. For more details on these concepts, see Appendix VI to *Radiological Laboratory Sample Analysis Guide for Incidents of National Significance–Radionuclides in Water* (EPA 2008a).

II. RADIONUCLIDES

The list in Table 1 is specifically for an RDD event and the major (non-inclusive) dose-related radionuclides that might be released during the detonation of an IND. In the case of an IND, numerous short- and long-lived radionuclides will be present, requiring proper identification and quantification. Several of the radionuclides on the list have progeny that coexist with the parents. Thus, if ^{228}Th were to be found, ^{224}Ra also would be present (although it is not listed). Several different radionuclides may be present even if only one RDD is used.

Instruments available for screening should provide a consistent measure of sensitivity[6] to allow detection of as many radionuclides as possible. However, some radionuclides (depending on total activity levels) likely will evade detection with routine screening instrumentation (solid scintillators or gas detectors). Generally, those radionuclides that decay by electron capture, positron emission, or very-low beta particle emission (and no gamma emission) should be analyzed with radiochemical-specific methods to determine their presence. The radionuclides from Table 1 in this group are: ^{3}H, 99Tc, ^{125}I, ^{228}Ra, ^{241}Pu, and ^{106}Ru. However, it should be noted that if liquid scintillation is used as a screening technique, a measurable response to these radionuclides will occur.

Table 1. Radionuclides of Concern

\multicolumn{3}{c}{Alpha Emitters}			\multicolumn{3}{c}{Beta/Gamma Emitters}	
Radionuclide	Half-Life	Emission Type	Radionuclide	Half-Life	Emission Type
^{241}Am	432.6 y	α, γ, [X-ray]	^{227}Ac[2]	21.77 y	β, ψ
^{242}Cm	163 d	α	^{141}Ce[1]	32.51 d	β, ψ
^{243}Cm	29.1 y	α, γ	^{144}Ce[3]	284.9 d	β, ψ
^{244}Cm	18.10 y	α	^{57}Co[1]	271.7 d	ε, ψ, X-ray
^{237}Np	2.14×10^6 y	α, γ, [γ, X-ray]	^{60}Co[1]	5.271 y	β, ψ
^{210}Po [1]	138.4 d	α	^{134}Cs[1]	2.065 y	β, ψ
^{238}Pu	87.7 y	α, [γ, X-ray]	^{137}Cs[4]	30.07 y	β, ψ
^{239}Pu	2.41×10^4 y	α, [γ, X-ray]	^{3}H[1]	12.32 y	β only
^{240}Pu	6.56× 10^3 y	α, [γ, X-ray]	^{125}I[1]	59.40 d	ε, β, X-ray
^{226}Ra[2]	1.60×10^3 y	α, γ	^{129}I[2]	1.57×10^7 y	β, γ, X-ray
228Th[2]	1.912 y	α, γ	^{131}I[1]	8.021 d	β, γ
^{230}Th	7.538×10^4 y	α, γ	^{192}Ir[1]	73.83 d	β, γ
^{232}Th	1.405×10^{10} y	α	^{99}Mo[2]	65.94 h	β, γ
^{234}U	2.455 ×10^5 y	α	^{32}P[1]	14.26 d	β only
^{235}U	7.038×10^8 y	α, γ	^{103}Pd[1]	16.99 d	β, γ
^{238}U[3]	4.468×10^9 y	α	^{241}Pu	14.29 y	β, [α,γ]
U-Nat[3]	---	α	^{228}Ra[2]	5.75 y	β only
			^{103}Ru[2]	39.26 d	β, γ
			^{106}Ru[2]	373.6 d	β only, (13, γ from progeny)
			^{75}Se[1]	119.8 d	ε, γ
			^{89}Sr[1]	50.53 d	β only
			^{90}Sr[2]	28.79 y	β only
			^{99}Tc[1]	2.11×10^5 y	β only

Notes:
The half-lives of the nuclides are given in years (y), days (d) or hours (h).
[1] No radioactive progeny or progeny not analytically useful.
[2] Radioactive progeny with short half-lives, and the progeny may be used as part of the detection method for the parent.
[3] Radioactive progeny not used for quantification, only screening.
[4] Radioactive progeny used for quantification only, not screening.
Brackets [ ] indicate minor emission probability. If large quantities of these radionuclides are present, these minor emission modes may contribute significantly to any screening measurements made on the sample.

III. Discussion

The discussion section of the document is divided into five parts. Part A deals with sample screening and different instruments that are commonly used to make these measurements. This section also provides some insight into technical issues encountered when performing gross sample activity measurements when the radionuclide being measured is unknown.

Part B deals with the calibration of screening equipment and the effects on the calibration process as a function of the particle type emitted by the calibration source and its energy It also discusses the responses of different types of detectors and provides figures demonstrating detector and sample configurations that may be advantageous for screening of samples for gross activity.

Part C deals with the use of screening equipment for prioritizing samples when the radionuclide(s) present are known.

Part D discusses the MQO process, and Part E provides key recommendations for the laboratory in establishing a screening protocol for samples resulting from radiological incidents.

A. Sample Screening and Processing at the Laboratory

Guidance on using both the screening instrumentation and the radiation-specific detectors for emergency response sample screening is discussed in this section.

Gross Activity Measurement Instruments

If the sample screening process at the laboratory is organized properly, it can significantly improve the turnaround time for results and minimize risk of the spread of contamination in the laboratory, as well as the chance for sample cross-contamination.

Gross activity measurements can be made using two general types of instrument—a ratemeter or a scaler.

The ratemeter measures the radiation emission per unit time in real time, but not all instruments have a summation function that would allow total decays to be measured over a defined time period. The overall sensitivity and ability of these instruments to discriminate radiation types are generally low. Although these are portable instruments that are often used for general area surveys, for the purposes of this guide these instruments are used in a fixed geometry relative to the samples. These instruments have time constants whose duration can be changed so that an average response to general radiation measured is more easily determined. A shorter time constant display has more frequent readings with the subsequent result of a "jumpy" needle or scale display when activity levels are close to the background level. By increasing the time constant, these measurements are averaged out internally and the display becomes more constant. This is more of a benefit to the application where the sample and the detector are in a fixed juxtaposition. When using a ratemeter for assessing gross radiation levels, it will be necessary for the laboratory to establish a protocol to determine the measurement value when meter/display readings are not constant (e.g., average the values of the high- and low-meter readings during a 20-second observation).

The scaler measures individual events and records them during a specified time period. Instrument outputs are generally in terms of total counts. The assessment of the gross activity generally takes longer with the scaler than with the rate meter, but the interpretation of the values obtained is somewhat more definitive. Some of these instruments have modest energy discrimination capabilities. However, these capabilities are severely limited when a mixture of radionuclides of varied decay modes is present. Laboratories should have a protocol that describes how to use the gross count data obtained by these types of instruments.

Table 2 identifies general descriptions of gross activity measurement instruments and laboratory screening instruments that can be used for sample screening and specific emission types to which they are most sensitive.

Instrument Response Characteristic Determination

The first factor to consider when performing a sample survey is the actual response by the instrument to the potential radionuclides plus any decay progeny in the sample. Not only is the response of these instruments different for each type of radiation, but it may also vary in a complex way with respect to the energy of decay. A couple of examples that demonstrate these differences in response are:

- The response of a NaI(Tl) micro-R meter will be different for high-energy photons compared to low-energy photons (i.e., it over-responds to low-energy photons).
- A GM pancake detector will respond to both alpha and beta radiation. However, for equal activities of ^{32}P (beta-emitter) and ^{242}Cm (alpha-

emitter), the instrument will yield a greater response (i.e., higher counts per minute) from the betas of ^{32}P.
- An open-end GM detector will respond to both beta and gamma radiation. However:
  - The response to 10 nCi of ^{89}Sr ($E_{\beta max}$ at 1.49 MeV) will be greater than that for 10 nCi of ^{99}Tc ($E_{\beta max}$ at 0.294 MeV).
  - The response to 50 nCi of ^{137}Cs (gamma energy 0.662 MeV) will be smaller than that for 50 nCi of ^{57}Co (gamma energies at 0.136 and 0.122 MeV) because the lower energy gamma rays interact more favorably due to the photoelectric effect.

These examples illustrate that the type and energy of radiation, as well as branching ratios, abundance values, and other physical properties of the radionuclide and the detection system are significant factors in assessing the total activity of a sample during the screening process using survey meters when the exact types of radionuclides present are unknown. Radionuclide-specific detection parameters are explained in detail in Knoll.[7]

Table 2. Detectors Used for Gross Sample Screening

Type of Detector	Sensitive to:
Geiger-Mueller (GM) Detector [Ionizable Gas]	Gamma (X-rays)
Open-end GM Detector [Ionizable Gas]	Beta, Gamma, X-rays (some high activity alpha)
GM Pancake Style Detectors [Ionizable Gas]	Beta and Gamma (some high activity alpha)
Micro-R meters [NaI(Tl)]	Gamma and X-rays
Cylindrical Probe [NaI(Tl)]	Gamma and X-rays
Thin Window (Alpha Scintillator)	Alpha
Thin Window (Beta Scintillator)	Beta (low response to photons)
Dual Phosphor Detectors First Layer [ZnS] Second Layer [Organic]	Alpha and Beta
Portable Gamma Detectors [HPGe]	Gamma (X-rays)
Small Article Monitors [NaI(Tl)]	Gamma (X-rays)
Small Article Monitors [Organic Scintillator]	Beta and Gamma (X-rays)
Liquid Scintillation [Liquid Fluor]	Beta, Alpha, and Gamma

Crosstalk: Detector Responses to Radioactive Emissions

In addition to the individual particle energy providing a different response in a particular detector, one type of particle may yield a response indicative of another type of particle. This is particularly true with gross alpha-beta detection devices that rely on pulse size to determine whether an individual event represents an alpha, beta, or gamma detection.

One instance of this type of incorrect identification occurs with measurement of ^{241}Am using a gas proportional detector. Although ^{241}Am is principally an alpha-emitter, it also emits a low- energy photon at 59 keV. A photon of this low energy may yield a response in the beta channel because of the high probability of secondary interaction of scattered radiation with the instrument components (including electronics, detector casing, instrument housing) via the photoelectric and Compton effects. Thus, if the total activity of the ^{241}Am is high, an incorrect assumption regarding beta activity could be made.

Care must also be used to evaluate and interpret the results with respect to possible beta-to-alpha and alpha-to-beta crosstalk effects when screening air filters (or other solid materials) for gross alpha and beta activities by instruments using gas proportional counting. The type of effect depends on the instrument mode of operation, setup, voltage plateaus, and discriminator settings. For most modern gas proportional counting instruments, the mode of operation may include:

1. Simultaneous measurements of alpha and beta activities based on a single operating plateau and beta-to-alpha and alpha-to-beta discriminator settings; or
2. Independent analysis of alpha and beta plus alpha activities on two separate voltage plateaus.

The second mode of operation, for most practical purposes, eliminates the beta-to-alpha crosstalk effect. However, the alpha response on the beta plateau must be estimated and the beta results adjusted accordingly. The remainder of the discussion that follows here will address the simultaneous alpha- and beta-counting mode.

The instrument voltage discriminator setting[8] should be adjusted when operating in the simultaneous alpha and beta counting mode to maximize the alpha detector efficiency and minimize the beta-to-alpha response crosstalk. These settings should be established using a source with matrix characteristics similar to the samples received from the incident response since absorption of the alpha particles in the matrix will decrease the alpha energy available with a proportional decrease in the signal voltage for processing. Typically, nominal instrument settings can be established that allow for an acceptable alpha counting efficiency and a beta-to-alpha crosstalk of <0.1 %. However, depending on the sample matrix and instrument settings, the actual crosstalk value can vary widely from this value. For air filter matrices, the alpha detector efficiency may be as low as 5 to 10%, and the beta-to-alpha crosstalk may contribute significantly to this value.

When evaluating gross alpha and beta activity results of sample analyses for the purpose of sample prioritization (for subsequent radionuclide-specific analyses), it is important to consider the possible effect of the beta-to-alpha crosstalk on deciding if the instrument alpha results have been artificially increased. The beta-to-alpha crosstalk effect may be most important either during the initial phases of an incident (when the radionuclides of interest are unknown) or when the composition of the mixture of alpha- and beta-emitting radionuclides is known. For the latter case, the beta-to-alpha crosstalk effect should be addressed. This can be done, once the radionuclides have been identified, by performing instrument calibrations for crosstalk using the actual radionuclides of concern, and corrections can be made that are both accurate and of known uncertainty.

A general observation of the AALs for those alpha- and beta-emitting radionuclides identified in Table 1 indicates that the AALs for the beta-emitting nuclides are at least a factor of 500 or greater than for the alpha-emitting nuclides. For example, the 500 mrem AALs for ^{90}Sr and ^{137}Cs are 110 and 550 pCi on the air filter for a 68 m^3 air sample. For the same dose and volume sampled, the AALs for ^{241}Am and ^{239}Pu are 0.17 and 0.14 pCi. For gross screening sample prioritization, the AALs for the ^{90}Sr and ^{241}Am should be used. Note that when the actual betato-alpha crosstalk discrimination is 0.1%, the alpha response observed from ^{90}Sr activity at the AAL may be > 0.1 cpm. With an alpha detector efficiency of 10%, the reported activity would be near the gross alpha screening AAL. Therefore, when evaluating gross alpha results when the beta result is greater than 500 to 1,000 times the alpha result, care must be taken to avoid the false conclusion that the screening alpha AAL has been exceeded. When screening air filters that have a beta-emitting radionuclide whose AAL is greater than the ^{90}Sr AAL, the beta-to-alpha crosstalk effect may be greater (depending on the beta particle energy), and the gross alpha screening AAL may be artificially exceeded more often when the radionuclide beta activity is near its own AAL.

As an example, suppose ^{90}Sr at the 500 mrem AAL (110 pCi/m^3) had deposited on an air filter. The activity would be the sum of ^{90}Sr + ^{90}Y = 220 pCi/m^3 for a 68 m^3 sample[9] (a total activity of 3.32× 10^4 dpm). The measured beta activity for a 30% efficient detector would be

beta dpm = 0.3 × 3.32× 10^4 dpm beta = 9.96× 10^3 cpm.

The alpha response from beta-to-alpha crosstalk would be based on the crosstalk factor, which is relatively small (about 0.1%). Thus, the apparent alpha activity counted would be

cpm = 9.96×10^3 × 0.001 = 9.96 cpm.

Alpha background on a GPC will be small at ~ 0.05 cpm. Thus, with an alpha efficiency of 0.1 (10%), the net count rate for alpha would yield a calculated alpha activity of

alpha = (9.96-0.05) / (0.1) = 99.1 dpm = 44.7 pCi.

This would yield a false indication of alpha activity when none is present.

Using the same reasoning, example AALs can be applied to the evaluation of air filters with elevated alpha activity. The effects of alpha-to-beta crosstalk (versus beta-to-alpha crosstalk) can be calculated, and the potential impact on artificially exceeding the beta AALs can be determined.

When operating a gas proportional counter in the simultaneous alpha and beta counting mode, the initial adjustment of the voltage discriminators is intended to minimize the beta-to-alpha crosstalk. Crosstalk, however, is more dependent on the specific radionuclide present in the sample and its physical decay and emission properties, than on the instrument discriminator settings. Actual alpha-to-beta crosstalk can vary from less than 3% to more than 30%, depending on the radionuclide and other factors.

Alpha-to-beta crosstalk correction factors should be determined during the initial instrument efficiency calibrations. These factors can be useful in making corrections to the beta count rate, based on the alpha count rate, but only when the radionuclide present has been correctly identified and the instrument has been calibrated accordingly.

When performing gross screening analyses, however, where the radionuclide has not been identified and the instrument has not been appropriately calibrated, making a crosstalk correction based on the initial instrument calibration can result in significant errors in the measurement of the beta activity in the sample. Depending on the project MQOs and event circumstances, it may be preferable to make no crosstalk correction and to potentially overestimate the sample beta activity. Because the beta AALs are typically much higher than the alpha AALs, this overestimate should result in artificially exceeding the beta AALs only when the alpha activity is extremely elevated.

In the previous example for beta-to-alpha crosstalk, the 500-mrem AAL for ^{90}Sr is 110 pCi for a 68 m^3 air sample. An ^{241}Am activity of 2,200 pCi would be required to yield a beta channel signal that would correspond to the 110 pCi activity for ^{90}Sr, or nearly 13,000 times the ^{241}Am AAL. For other alpha-emitting radionuclides, the alpha activity required to cause this beta AAL to be artificially exceeded could be greater than 100,000 times the AAL of that other radionuclide.

In these unusual cases, the apparent beta activity should be confirmed by an appropriate technique, such as recounting the sample with an alpha-attenuating barrier in place and comparing the beta count rates from the two analyses. For screening analyses, however, these techniques assist only in estimating the degree of bias in the results, and do not correct for all sources of crosstalk.

This effect can be illustrated by calculating the quantity of alpha activity from ^{241}Am that would yield an indication of beta activity at the AAL for ^{90}Sr. Given the 10^{-6} risk AAL for ^{90}Sr of 0.29 pCi/m^3 and an assumed sampled air volume of 68 m^3:

- A beta activity from (^{90}Sr + ^{90}Y) on a filter at the AAL would be approximately 88 dpm;
- The beta counts recorded (with a detector efficiency of 30%) would be ~26 cpm beta; and
- A normal beta background of 1 cpm yields a net beta count rate of ~25 cpm.

Assuming 30% alpha-to-beta crosstalk and 10% counting efficiency for ^{241}Am, the alpha activity required to produce alpha-to-beta crosstalk equivalent to the ^{90}Sr AAL would be

Alpha activity = 26/(0.3 ×0. 1 ×2.22), or approximately 390 pCi.

Thus, an activity of ^{241}Am of 390 pCi can cause an *apparent* beta activity equivalent to the AAL of ^{90}Sr even when there is none present.

Detector Background

A second factor to consider during sample screening is the background. Background can be divided into the categories of instrument (intrinsic or electronic), environmental (laboratory location), and sample container/sample. These should be minimized when possible to achieve the best signal to background ratio for the sample. As will be shown further on in this document, reduction of background is one of the most important limiting factors for detection of low level sample activity during the screening process.

Some examples of potential background concerns are:

- Proximity of one screening instrument to another when samples or groups of samples contain enough activity to have an impact on a neighboring instrument.
- Presence of radionuclides with multiple emissions that can be detected by the instrument.

Since the level of background is crucial to the measurement, the shielding of the detector is an important consideration.

Sample Geometry

The third factor that should be considered when using survey meters is the consistency of the sample-to-detector geometry. The method of calibration of the survey meter and the method used to screen samples using the survey meter should match as closely as possible to obtain the best estimate of absolute activity in the samples.

Finally, sample self-absorption should be evaluated when assessing the results of sample screening. This effect is most critical with alpha- or beta-emitters, but for low-energy photon- emitters it also will be a contributing factor to misidentification of particles. The loss of particle energy as it travels through the sample medium will cause it to yield a smaller ionization pulse in the detection device. As described earlier, this can register a false count for the wrong type of emitted particle.

Each of these three factors will be considered in the sections below that address the calibration of screening detection equipment.

Laboratory Instruments

Hand-held devices are not the only types of instrumentation that can be used for performing a gross radiological screen on a sample. Consideration should also be given to using three mainstays of the radiochemical laboratory for screening analyses. Gas proportional counters (GPC), NaI(Tl) detectors, and liquid scintillation counters (LSC) normally are used for radionuclide-specific analyses, and in such applications radiochemical purity of the sample test source (STS) is imperative. These instruments can be used to assess total activity as well. This may require a modification or re-configuration of laboratory instrumentation to dedicate some portion of the laboratory resources to emergency response rapid screening.

B. Calibration of Instrumentation for Screening Analyses

Detector Type

Examples of different types of gross screening survey meters and laboratory screening instruments are summarized here:

- Gross Alpha
    - ZnS(Ag) scintillation detector with a thin aluminum or Mylar™ window
    - Open-end GM detector
    - Gas-filled pancake probe with a thin window
- Gross Beta
    - Plastic organic scintillator with a thin aluminum or Mylar window
    - Gas-filled GM detector (with slide-window allowing gamma detection in the presence of beta)
- Gross Gamma
    - Gas-filled GM detector
    - Sodium iodide (NaI[Tl]) or cesium iodide (CsI[Tl]) detector (well or flat type crystal) with scaler for open discrimination counting
    - Micro-R meter using NaI(Tl) or CsI(Tl) detector
    - HPGe detector (may be flat or well type) set for gross counts using summation of all channels

Table 3. Radionuclides Spanning the Energy-Calibration Range

Radionuclide	^{57}Co	^{60}Co	^{137}Cs	^{99}Tc	^{90}Sr/^{90}Y	^{230}Th[1]	^{241}Am
Emission Type	γ	γ	γ	β_{max}	β_{max}	α	α
Energy, MeV	0.122, 0.136	1.173, 1.332	0.662	0.29	0.545, 2.28[2]	4.69	5.49

[1] This is the primary alpha for thorium; thorium has progeny that emit alphas as well.
[2] This energy belongs to ^{90}Y, which is in secular equilibrium with the ^{90}Sr.

It would not be practical to maintain calibrations for each of the radionuclides, or mixtures of radionuclides, shown previously in Table 1. However, a straightforward process can be performed to relate the response of each detector to decay particle energy. While the measurements are not as precise as more extensive laboratory measurements, it allows increased accuracy for a longer list of radionuclides when making an estimate of the total activity. This can be accomplished by selecting at least two (but preferably three or more) radionuclides that emit characteristic decay particles with distinct energies that span the usable range of the instrument. Table 3 identifies a list of radionuclides that can be obtained as standards for calibration of detector energy. Their emissions and energies for calibration are also included.

Next, the net instrument response for each of the radionuclides is measured in a standard configuration (i.e., a "geometry": matched quantities of sample, containers, and position relative to the active volume of the detector). For each type of decay particle and geometry, instrument response should be plotted against the average decay energy[10] of the particle emitted. Using these data, a table of response factors (i.e., efficiencies) is prepared that correlates to each of the radionuclides in Table 1 based on decay type and respective average decay energy. An example of this application can be seen in Figure 1, which shows the energy response to different energy gamma radiation for a halogen quenched GM detector, and in Figure 2 for a NaI(Tl) detector. Note the significant, relative effect that using the GM shield has on the detection of the lower- energy versus the high-energy gamma emitters. This also can be used in a qualitative sense to assess the overall energy profile of the gamma emitters.

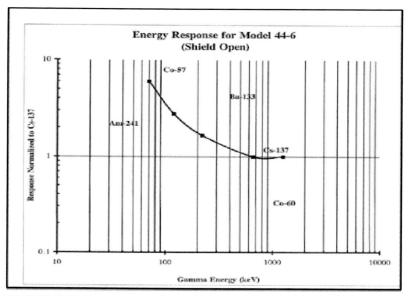

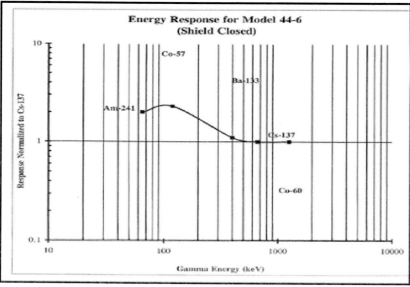

Figure 1. Halogen-Quenched GM Detector Response to Gamma Radiation (A) with Shield Open (B) with Shield Closed.

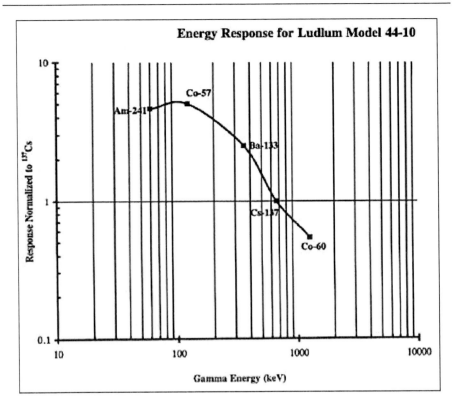

Figure 2. Gamma Energy Response for a Na(Tl) detector.

The maximum in detector response for the commonly used NaI(Tl) detector is about 100 keV (see Figure 2). For a comparable sized CsI detector, the response would be more efficient overall, and the maximum in the efficiency curve would be at a slightly higher energy. This is due to the difference in physical properties of the CsI(Tl) crystal material.

The response for an alpha-beta survey meter,[11] using a halogen quench fill gas and a thin mica window pancake probe, may have the following characteristics:

- Efficiency (2-pi geometry): 5%–^{14}C; 22%–^{90}Sr/^{90}Y; 1 9%–^{99}Tc; 32%–^{32}P; 1 5%–^{239}Pu
- Sensitivity: Typically 3,300 cpm/mR/h (^{137}Cs gamma)
- Energy Response: Energy dependent

From these few examples, it can be seen that the response of a survey instrument to different types and energies of radiation is a complex function of not only the radiation emitted but also of the survey instrument used.

Geometry

The relative geometry of sample to screening instrument and shielding can take on several different configurations. It is very important to ensure that the sample measurement matches the calibration geometry. Some of the considerations that will affect the optimal configuration of sample to detection device are:

- Shielding (detector). The detection capability of the screening method will be optimized by shielding the detector to reduce ambient background and minimize response to external sources of radiation. The detector and detector shielding configurations should remain fixed so that the background count rate is reasonably constant.
- Shielding (container). The sample container material can be made of glass, polyethylene, Teflon, or other non-reactive material. The effect that these different materials have on shielding the radioactive emissions from the detector varies with particle type and energy. Also, the thickness of the container walls can increase the average distance of the centerof-activity of the sample to the detector. Both of these sample container characteristics can affect the net screening result.
- Volume/shape/density. The sample volume must be consistent with gross measurements made during the calibration of the screening equipment so that the relative configuration of sample-to-detector is maintained. Thus, it is important that the sample container be virtually identical to the container used for calibration purposes. Sample density (or for solids, the degree of compaction) has a significant effect on the potential self-shielding of the sample from the detector. The mass of the calibration source and the sample should be relatively close in value to achieve consistent configuration.
- The figure of merit[12] (FOM) for the configuration of the shielding may need to be optimized (i.e., a larger FOM is better). For example, it may be advantageous to have a relatively large shielded volume with the sample centrally located, versus a shielded volume that exactly fits the sample geometry.

- Location of the sensitive detection area in the screening equipment. The manufacturer's detailed diagram for the specific model of screening equipment should be available so that the optimum position of the detector with the sample can be achieved (See Figure 3).
- Size and shape of the detector with respect to the sample geometry. The sample shape and detector juxtapositioning can have significant effects on the measurement. One measure of this is the FOM.

Table 4. Response and Figure of Merit for ^{60}Co and ^{137}Cs with Different NaI(Tl) Detector Configurations

Radionuclide	Activity pCi/L	Net cpm			Figure of Merit		
		1"×1"	2"×2"	3"×3"	1"×1"	2"×2"	3"×3"
Background (BO)	—	2.80×10²	1.65×10³	2.4×10³	—	—	—
^{137}Cs (BO)	5.038×10⁵	2.22×10³	1.34×10⁴	1.3×10⁴	6.93×10⁻⁸	4.28×10⁻⁷	2.77×10⁻⁷
^{60}Co (BO)	3.317×10⁴	5.5×10²	9.5×10²	7.5×10²	1.67×10⁻⁷	4.97×10⁻⁷	2.13×10⁻⁷
Background (SO)	—	2.7×10²	1.1×10³	1.75×10³	—	—	—
^{137}Cs (SO)	5.038×10⁵	5.03×10³	1.66×10⁴	1.78×10⁴	3.69×10⁻⁷	9.87×10⁻⁷	7.13×10⁻⁷
^{137}Cs (SO)	5.038×10⁴	3.1×10²	9.0×10²	6.0×10²	1.4×10⁻⁷	2.90×10⁻⁹	8.1×10⁻⁸
^{137}Cs (SO)	1.242×10⁴	1.5×10²	3.0×10²	1.2×10²	5.40×10⁻⁷	5.30×10⁻⁷	5.33×10⁻⁸
^{60}Co (SO)	3.317×10⁴	5.8×10²	2.2×10³	1.5×10³	1.13×10⁻⁶	3.99×10⁻⁶	1.17×10–6

Notes:
SO = Side Orientation (see Figure 3)
BO = Bottom Orientation (see Figure 3)
Example Calculation: For the Cs bottom orientation (BO) and the 1 "×1" detector
FOM = [net cpm/pCi/L]2/[Background] = [2.22×10³/5.038×10⁵]²/(2.8× 10²) = 6.93×10⁻⁸

An example illustrating the effects of the size and shape of the detector on the FOM can be seen in Table 4, which identifies some data taken using NaI(Tl) detectors of various sizes (none of these were well detectors). The sample container was a 1 liter plastic bottle. The data were recorded using a detector and shielding as shown in Figure 3. The configuration of the detector and shielding actually used in this case was not optimal: In the bottom orientation position, the detector is partially unshielded, and the flat surface of the NaI(Tl) detector can is facing the sample bottle. Figure 3 also shows the side orientation where again the detector is partially unshielded, and the curved detector cover is parallel to the sample. Also, note the actual position of the detector crystal in both cases. It is clear in either case, however, that detector size and positioning with respect to the sample will have a significant effect on the measurement sensitivity (based on the FOM).

The data indicate that the biggest detector volume does not always give the highest count rate, nor does it always yield the highest value FOM. Thus, it is imperative that the detection equipment used be assessed in a similar fashion to determine which screening equipment is best suited for each combination of matrix and geometry. Two factors to be considered in determining this are:

- Location of the mean sample activity relative to the location of the detector, and
- Shielding (covering) of the screening equipment.

There are different considerations for samples that need to be screened for gamma radiation. An example is using a NaI(Tl) well detector. Many different sample types can be accommodated into this well for screening purposes. For example, a 47-mm air particulate filter may be rolled and inserted into a container, such that the container will fit reproducibly into the well of the NaI(Tl) detector, improving overall efficiency for detection. When doing this, care must be taken to avoid contaminating the detector. That specific geometry for calibrating this style of detector can be accommodated by most laboratories.

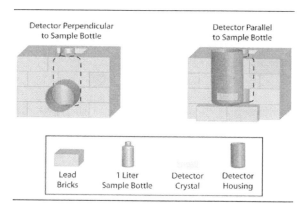

Figure 3. Shown Without Bricks Covering Top of Shielded Geometry. NaI(Tl) Detector Example.

Radiological Laboratory Sample Screening Analysis Guide ...   107

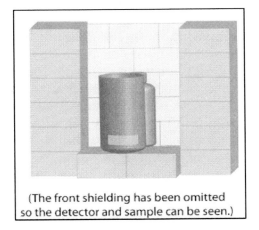

Figure 4. An Improved Orientation for Shielding. Active Detector Area Within Shielding.

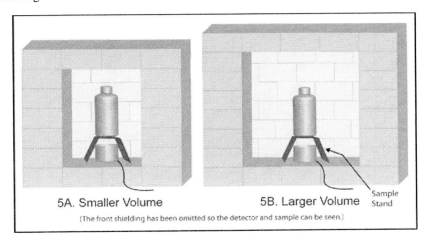

Figure 5. Relative Size of Shielded Volume.

Figure 4 shows another way to configure the detector and the sample bottle to achieve a better FOM for the measurement. In this configuration, the active area of the NaI(Tl) detector is inside the shielding and thus has a lower net background from room and ambient background contributions.

Figure 5 shows two different configurations of shielding with respect to the detector that will provide different backgrounds. Note that the thickness of the shielding walls is the same but that the internal cavity in which the detector is held is larger in Figure 5B. The larger volume ultimately leads to a better

FOM since any Compton scattering from the shielding in 5B will impinge to a lesser degree on the detector than in 5A solely due to distance. In Figure 5B, a sample stand has been added to put the sample in the middle of the shielded volume, and the detector has been raised slightly to yield the same orientation as in 5A, thus maintaining the same detector efficiency.

Figure 6 shows a configuration for a pancake-style screening instrument (could be gross alpha- beta or beta-gamma). The air particulate filter is slid into place beneath the detector, which is maintained in a fixed position using a small stand. The presence of shielding allows reduction in background for the detector and for the sample, and provides a fixed geometry for consistent results.

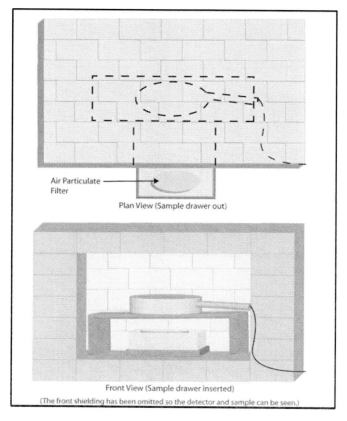

Figure 6. Sample Shielding and Detector Orientations for Gross Screening of Air Particulate Filters Using an Alpha/Beta Pancake Detector.

Crosstalk, Dead-Band, and Self-Absorption Factors

The degree of crosstalk as determined under routine instrument calibration conditions may not be significant. However, when the activity being measured is two and three orders of magnitude greater than normal sample test sources, crosstalk that was once obscured in the background may provide a signal that is indicative of a particle type that is absent. Thus, it is important to challenge the screening instrumentation with standards of high activity so that the level of crosstalk can be assessed. One such application involves GPC systems that are simultaneously counting gross alpha and beta activity. An assessment of crosstalk should be made in the beta channel response when the alpha activity is large compared to the beta activity and compared with the same beta activity response with no corresponding alpha activity. The inverse assessment should also be made. These measurements may lead the lab to apply a "dead band" between the lower level beta and upper level alpha discriminator settings that normally would not be used. This dead band would minimize the crosstalk, but would also lower the efficiency for both types of particles. Thus, the use of a dead band should be used judiciously to avoid abnormally long count times when screening time is at a premium.

It should also be recognized that elevated activity of radionuclides that decay only by beta emission may result in counts above background when using a sodium iodide detector for gross count assessment (e.g., as when using a small article monitor). The *bremsstrahlung* radiation, emitted as a result of the beta interaction with matter, yields low-energy photons that produce a signal in the sodium iodide detector.

Self-absorption factors are significant for alpha- and beta-emitters. Determining how sample mass affects the efficiency of detection can be estimated using calibration sources and absorbing materials of known areal density (measured in units of mg/cm^2) placed between the sample and the detector. This intervening material would simulate the sample mass when the sample is not ideal (i.e., the sample is not "massless" and will absorb some of the contained radiation). This mass attenuation correction for self-absorption is similar to determining unknown beta particle energy using the Feather Method.[13] For alpha particles, this may mean using a thin film of aluminized Mylar, while for betas, varied thicknesses of aluminum metal may be used. The areal density effect for each detector should be semi-quantitatively identified so that estimates of activity correction can be made when samples of observable mass are measured using detection techniques such as GPC.

Table 5. Screening Instrument Conversion Factor Based on Sample Analysis of a 1-Liter Sample Geometry

Sample	Screening Value, (mR/h)/L	Radionuclide-Specific Analysis Results, ^{137}Cs µCi/L	Conversion Factor [µCi]/(mR/h)	Estimated Conversion Factor Uncertainty [µCi]/(mR/h) [1]
Background	2	—	—	—
Sample 1	55	1,601	30.2	—
Sample 2	78	2,005	26.4	—
Sample 3	41	1,448	37.1	—
Average Conversion Factor: 31 ± 5				

[1] The method used to estimate the screening equipment uncertainty must be decided upon by the laboratory. The column is included here so that it is clear that this should be one aspect of this process.

Final Instrument Calibration and Method Validation

Once the detectors to be used for screening have been selected and the considerations for sample to detector configuration and efficiency have been assessed, a method should be written. The method should incorporate the laboratory's best estimate of the potential geometries and plausible radionuclides into the procedure. Specific instructions regarding the receiving and storing of the samples, recording of data, and sample aliquanting for particle-specific screening should be included in this method. Once the method is written, a method validation process that follows the *Method Validation Requirements for Qualifying Methods Used by Radiological Laboratories Participating in Incident Response Activities* (EPA, 2009b) should be followed. The method validation process requires the use of proficiency test samples to validate the detector response to achieve the MQOs established for the project or by the laboratory. Once the method has been validated, the procedure should be implemented routinely for sample processing by all staff members, which will reinforce training on the procedure.

C. Calibration of Screening Instruments when Radionuclide Identities are Known

Screening equipment that is calibrated for overall response to decay particles will have its accuracy challenged if the radionuclide in the sample to be measured has a different particle or energy.

During the initial phases of an emergency, before the identity of the radionuclide(s) associated with the event has been established, a response factor for the screening equipment presumably will be based on a single radionuclide, such as ^{137}Cs. As the radiological event progresses, the radionuclide(s) associated with the event should be identified. For example, if ^{192}Ir is identified, the factor used to convert cpm/sample to pCi/sample should be changed so that the screening equipment more accurately characterizes the sample activity level, and the laboratory will be able to characterize the activity of the samples more accurately. This change in the response factor can be implemented in several ways:

1. The laboratory has already established a response factor on the screening equipment for this radionuclide in this geometry. In this case, receipt instructions need to be updated to include the identity of the radionuclide(s) of concern. For example, consider a beta/gamma survey meter that has been calibrated with a ^{137}Cs source that had a measured response factor for a 1-L liquid sample of 5.1×10^{-4} mR/h per pCi. This factor has been entered into the electronic database for the meter used (identified by serial number). Knowing now that the radionuclide of interest is ^{192}Ir, with a response factor of 2.8×10^{-4} mR/h per pCi, this response factor should replace the ^{137}Cs value currently present in the electronic database.[14] This change will identify more accurately the activity based on gross screening measurements.

2. The laboratory has performed an energy response factor curve for the screening equipment and can interpolate the curve for the effective mean emission energy of the radionuclide present in the sample. For example, this method is demonstrated in Figure 7 for a 47-mm air particulate filter using simulated data.

   In Figure 7, the radionuclide energies represented are approximately one-third of $E_{\beta max}$. Thus, as an example, the effective beta particle energy for ^{14}C is 156 keV/3 = 55 keV.[15]

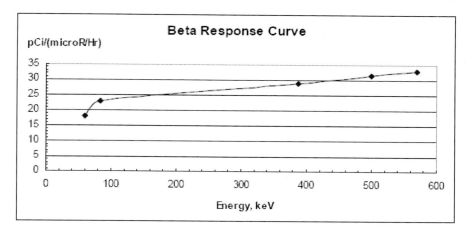

Figure 7. Survey Meter 12345 Energy Open Window Response Curve for Beta Emitters.

3. The laboratory may take a sample that is to be analyzed, and determine a conversion factor based on a comparison of the screening value and radionuclide-specific analysis results. In this case, it would likely be best to take an average conversion factor from several samples to ensure the most accurate representation of the factor. This is because the factor can be affected by non-uniform distribution in the sample. Consequently, the laboratory should consider the potential for significant uncertainty in this conversion factor, which may be estimated by the standard deviation of the individual measurements used to calculate the average conversion factor.

As an example, a data table like the one below could be constructed. Note that the information shown is not based on actual data but is used for illustrative purposes only.

In this example, the samples have already been screened using a micro-R meter. The samples are then analyzed using a radionuclide-specific method, and the values obtained are specifically for ^{137}Cs. The final analytical values for the samples are divided by the original exposure rate measurements to obtain a conversion factor for the radionuclide contained in the event-specific samples. The average conversion factor and the associated uncertainty estimate are rounded to the appropriate number of significant digits. In this case, the conversion factor would allow the laboratory to estimate the concentration of ^{137}Cs in the subsequent samples, based on the micro- R meter screening results. This simplified example uses a single radionuclide with no ingrowth considerations. In cases where one or more radioactive progeny may be present, care must be taken to ensure that the screening conditions, especially the degree of progeny ingrowth, are reasonably consistent. In all cases, the counting geometry for sample screening should be as consistent as possible.

During the latter phases of an event (when the radionuclide content of the samples is expected at the 10^{-4} risk level for air filters and the maximum contaminant levels for drinking water), the screening of lower activity samples may be performed using a different technique. For example, if both alpha- and beta-emitters are present, rather than using GPC to screen the samples for both alpha- and beta-emitters simultaneously, it may be advantageous to perform each screen separately and extend the count time to ensure better discrimination between those samples where analysis is required immediately and those that may be delayed.

D. Measurement Quality Objectives (MQOs) for the Screening Process

Screening of samples as they arrive significantly impacts the laboratory's decisions about which samples to analyze first. The IC should have decided how the samples are to be prioritized and communicated this to the laboratory. The laboratory may confidently screen these samples for gross activity so that they can be processed in a timely fashion based on the needs of the incident.

General guidance on how to establish an MQO for the required method uncertainty can be found in MARLAP (2004) and specifically for radionuclides in water (EPA 2008a, Appendix VI). Additional MQOs for screening should be established by the laboratory based on the type of instrumentation available.

In order to illustrate the typical decisions and actions to be taken by a laboratory for calibration and gross sample screening, three examples using theoretical samples and measurement results are provided in Appendices I-III. These examples demonstrate an acceptable method for the calibration of instruments and measurement of samples, but each example is one of several different possible variations of calibration and measurement techniques. The examples here should not be construed as limiting.

The first scenario (Appendix I) illustrates how a laboratory *may prepare* its screening equipment to be ready to receive samples from a radiological incident. The instrumentation and standards used are limited to what is available to the laboratory, which demonstrates how some basic planning can assist in being prepared for such an event. In the second scenario (Appendix II), the same laboratory *has received* samples from a radiological transportation accident and has been asked to rapidly assess the spread and degree of contamination. The calibration of the screening equipment is optimized to assess the contamination levels in the samples that have been sent. The third scenario (Appendix III) discusses how instrument calibrations may be adjusted during the latter phases of an event when the radionuclide(s) identity(ies) is (are) known. In this instance, the screening process will be looking at lower overall activity in the samples so that re- calibration with the same radionuclide will enhance the detection capability of the screening equipment.

E. Key Recommendations

Laboratories should be prepared for potential radiological events where large numbers of samples at much higher activity concentrations than normal arrive suddenly. To assist laboratory personnel in promptly receiving, prioritizing, and analyzing samples, the following is a summary of the key recommendations for sample screening:

- Screening equipment should be calibrated with traceable sources that match geometries for anticipated emergency response samples.
- These calibrations should have associated direct reading conversion factors for ease of reporting results in the appropriate units.
- Laboratories should have written procedures (or instructions) for the process of screening emergency response samples.
- Shielding for the screening equipment should be configured to maximize the signal to background ratio, providing the analyst with smaller uncertainties of the measurement.
- A plan that provides for the calibration adjustment of the screening equipment based on the incident radionuclide(s) should be prepared for that time when the activities are much lower, and better discrimination between lower level activities will be required.

APPENDIX I. SCREENING INSTRUMENTATION INITIAL CALIBRATION

> It is assumed that laboratories will have properly calibrated their instrumentation prior to an event. The data provided in the following three scenarios (Appendices I–III) are used for illustrative purposes only. Each laboratory should consider using the general techniques modeled here for its laboratory-specific methods to be used for sample screening. In addition, uncertainty values have been omitted from these examples. For actual calibration and screening, uncertainties should always be included in the expression of the final results.

Background

ABG Laboratory, Inc., has decided to set aside certain instruments for radiological events where sample gross screening will be necessary. A GM pancake detector and an old 3"×3", planar, NaI(Tl) detector with a scaler have separate, shielded geometries for samples of 47-mm filters, a 1 -L liquid, and a 250-g solids container. The equipment is to be located near the sample-receiving area of the laboratory facility. Once the equipment is set up, the laboratory staff performs background counts on the instruments while waiting for the new calibration sources to arrive. The calibration sources are ^{99}Tc, ^{90}Sr, ^{241}Am, ^{57}Co, ^{230}Th, and ^{60}Co. Each source has been ordered for each geometry identified above and is traceable to a national standards body, such as the National Institute of Standards and Technology in the United States.

Discussion

The NaI(Tl) detector was set up to accumulate total counts in a two-minute count. The GM pancake detector was set up in rate mode for cpm. The following table identifies the detector background and response from the standards for each of the instruments.

Table 6. Calibration Data for Screening Instrument Response

Detector	Radionuclide Source	Total Background*	Activity pCi	Air Filter, Net Counts	250 g Can, Net Counts	1 L Bottle, Net Counts
NaI(Tl)	^{57}Co	5,840 cpm	8.0×10^5	8.88×10^4	5.33×10^4	2.13×10^4
			2.0×10^5	2.22×10^4	1.33×10^4	5.33×10^3
	^{60}Co	5,840 cpm	8.0×10^5	1.24×10^5	9.59×10^4	3.66×10^4
			2.0×10^5	3.11×10^4	2.40×10^4	8.88×10^3
GM				Air Filter, Net cpm	250 g Can (open) Net cpm	1 L bottle (closed, side measurement) Net cpm
Alpha	^{241}Am	0.05 cpm	40	8	0.18	0
			10	2	0.04	0
	^{232}Th	0.05 cpm	32	6.5	0.14	0
			8.0	1.6	0.04	0
Beta	^{99}Tc	0.8 cpm	4.5×10^3	509	10	0.1
			1.2×10^3	136	2.5	0.03
	^{90}Sr	0.8 cpm	300	133	67	20
			80	35.5	18	5.3

* For the sodium iodide detector, background counts were summed over the energy range of 50 to 2500 keV. For the GM detector, the background represents an average measurement performed at several times of the day. Each instrument background measurement was made using an empty sample container in the position for sample measurement, and the sample plus detector were shielded with 4" of lead brick.

The laboratory staff has made separate calibration factors for low- and high-energy gamma-ray emitters. Similarly, for the ^{90}Sr and ^{99}Tc, the efficiency of detection of the ^{90}Sr is much better due to a smaller degree of self-absorption in the sample and better penetration of the GM detector beta shield when used. The response factors for both the ^{241}Am and the ^{232}Th are the same. The laboratory staff has made the following response factor table for its instruments:

The response factors in the table are calculated as follows:

$$RF = \frac{\text{Source pCi}}{(\text{net cpm})\cdot \text{Abundance Factor}}$$

Table 7. Response Factors (RF) for Radionuclides with Respective Screening Equipment

Radionuclide	Energy, keV	Abundance Factor*	Air Filter, pCi/cpm	Open Tuna Can, pCi/cpm	Bottle, pCi/cpm
NaI(Tl) Detector, Gamma					
^{57}Co	121, 135	1.003	17.5	29.9	74.8
^{60}Co	1,173; 1,332	2.0	6.45	8.34	22.5
GM Detector, Alpha					
^{241}Am	5,449; 5,440	1.0	5.01	2.25×10^2	1.5×10^5
GM Detector, Beta					
^{99}Tc,	210	1.0	8.83	4.50×10^2	4.50×10^4
^{90}Sr	546; 2,280	2.0	1.13	2.25	7.5

*The abundance factor is the number of particles that are produced per decay of the radionuclide and can be detected by the detector listed. The value for ^{60}Co is 2.0 since it yields two gamma rays for each decay (the gamma rays are in full coincidence). For ^{90}Sr, the value is 2.0 since it is in secular equilibrium with its progeny ^{90}Y, also a beta- emitter.

Thus, for the air filter geometry on the NaI(Tl) detector for ^{60}Co:

$$\text{RF}_{\text{Co-60}} = \frac{8 \times 10^5}{(1.24 \times 10^5 \text{ counts}/2 \text{ min}) \cdot 2.0} = 6.45 \text{ pCi/cpm}$$

APPENDIX II. RADIOLOGICAL EVENT SCREENING FOR ^{241}AM

Background

The date is November 15, and steady winds from the northwest at about 20-2 5 mph are expected through tomorrow. A truck is carrying used ^{99m}Tc generators[16] and [241]Am smoke detectors (as the bulk of its shipment, but other radioactive waste materials of smaller volume were on board). The truck overturns and slides into a rock embankment, bursting into flames along a small two- lane highway between towns, and burns down to the tires. Air sampling equipment has been stationed in several locations in both towns and along several roadsides. Air samples are expected to arrive at the laboratory by 1800 hours this evening (it is currently 1300 hours). Additionally, several hundred soil and crop samples are expected over the next week so that the plume can be tracked.

The IC has requested that the highest activity samples be identified and analyzed first so that the recovery phase can focus on:

- Determining how much material has become airborne, and
- Cleaning up high activity areas first to remove the bulk of the source term.

Discussion

ABG Laboratory, Inc., has been contacted and told to expect the samples shortly. It will be using the calibrations it has made for its screening equipment to accommodate the influx of samples.

Table 8. Gross Screening Measurement Results from Transportation Incident

Sample	Alpha GM Detector, cpm	α Gross Screening Estimate, pCi	Beta Open Window Probe, cpm	β Gross Screening Estimate, pCi	NaI(Tl) Detector, cpm	γ Gross Screening Estimate, pCi
Air Filter-1	4.7	23.3	1.77	8.57	5,750	-1,580
Air Filter-2	0.085	0.175	**5.82**	44.3	5,900	1,050
Air Filter-3	0.10	0.25	**2.88**	18.4	5,880	700
Soil-1	0.550	113	1.46	297	6,050	**1,750**
Soil-2	0.16	24.8	3.9	**1,395**	8,120	**19,000**
Soil-3	0.07	4.5	0.7	-45	6,000	1,330

Table 8 identifies the sample activity measured for each of the matrices received at ~1800 hours. Knowing the truck's cargo makes use of the calibration factors straightforward. The air particulate filters have been transmitted in glassine envelopes, and the soil samples were stored in solids (tuna) can geometry with a removable lid. The laboratory has verified that these geometries match the geometries it used for its gross screening calibration of the instruments.

The spreadsheet it is using has the following equations for the analysis:

- Air Filters
    - Gross Alpha Activity = (meter reading, cpm − 0.05, cpm)×(5.01pCi/cpm)
    - Gross Beta activity = (meter reading, cpm − 0.8, cpm)×(8.83 pCi/cpm)
    - Gross Gamma Activity[17] = (Total counts − 5,840 cpm)x(17.5 pCi/cpm)
- Solids Can
    - Gross Alpha Activity[18] = (meter reading, cpm − 0.05, cpm)×(225pCi/cpm)
    - Gross Beta activity = (meter reading, cpm − 0.8, cpm)×(450 pCi/cpm) o Gross Gamma Activity = (Total counts − 5,840 cpm)×(8.34 pCi/cpm)

The laboratory reports back to the IC that the sample results, bolded above, have the highest concentrations based on gross screening results, and the analyses for ^{241}Am and ^{99}Tc are in progress. The laboratory supervisor queues the samples according to activity. The highest-activity samples are to be analyzed first. The supervisor also notifies the separations chemists about the levels of activity they will find in these samples.

The laboratory protocol has established a limit of 100 pCi per aliquant. Normally, the sample size processed is 2.0 g. However, for Soil-2, there is 250 g of sample, and in order to be less than 100 pCi, only 1.0–1.3 g of sample will be aliquanted for this analysis.[19]

APPENDIX III. SCREENING INSTRUMENTATION RESPONSE CORRECTED FOR DIFFERENT RADIONUCLIDE

Background

A suspected terrorist event involving explosive devices has occurred. Several different radioactive materials suppliers have reported thefts of large quantities of radionuclides in the past three months. The missing radionuclides were ^{210}Po and ^{192}Ir. Preliminary evidence from the scene of the incident identified the presence of radioactive materials. It is suspected that the materials that were reported missing are related to this event.

Radiochemistry Analysts of America has been contacted to screen, then analyze about 200 samples a day for ^{192}Ir and ^{210}Po, and any other radionuclides that may be present. The samples will be air particulate filters (47 mm) and soil (~0.200 kg). It is Day 1 at 1100 hours, and the first sample shipment will arrive at 0600 hours on Day 2. The IC has indicated that the sample priority is to analyze those samples with the highest activity first. The laboratory has neither a ^{210}Po nor a ^{192}Ir source/standard.

Discussion

The laboratory has selected a NaI(Tl) well detector to screen the air particulate filter samples for the ^{192}Ir. Its current calibration factor used ^{60}Co, but it has a response curve based on energy as shown below.[20]

The average energy[21] of the ^{192}Ir is approximately 390 keV. This corresponds to a factor of 1.30×10^4 pCi/(net cpm) as estimated using the curve in Figure 8. A similar curve was made for the solid geometry and a response factor of 6.1×10^3 pCi/(net cpm) for ^{192}Ir was estimated.

The laboratory staff is using a GM pancake-style detector for alpha screening of the samples. The corresponding response factors for alpha particles are:

Air filter	Solid, 200 g
5.01	2.25×10^2

The following day, several hundred samples are received, and the screening process begins. An example dataset is shown below:

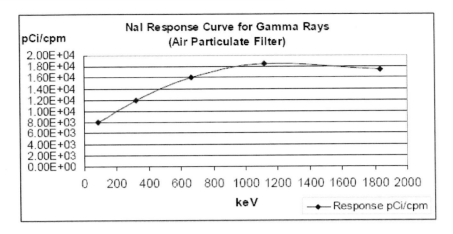

Figure 8. Gamma Energy Response Curve for a NaI(Tl) Detector.

Table 9. Results of Screening Measurement Using Adjusted Response

Sample ID	Air Filter 1	Air Filter 2	Air Filter 3	Soil Sample 1	Soil Sample 2	Background
NaI(Tl), cpm	4,630	4,550	4,480	6,100	4,700	4,500
GM Detector, cpm	2.80	1.56	0.23	0.13	0.14	0.12
Screening Results							
Gross gamma, pCi	1.7×10^6	6.5×10^5	-2.6×10^5	9.8×10^7	1.2×10^6	–
Gross alpha, pCi	13.43	7.21	0.55	2.25	4.5	–

Based on the results of these screening measurements, air filter 1 and soil 1 have the highest activities and should be analyzed first for ^{192}Ir and ^{210}Po.

APPENDIX IV. ADDITIONAL SOURCES AND REFERENCES

[1] U.S. Environmental Protection Agency (EPA). (1992). *Manual of Protective Action Guides and Protective Actions for Nuclear Incidents.* Washington, DC. EPA 400-R-92-001, May. Available at: www.epa.gov/rpdweb00/rert/pags.html.

[2] U.S. Environmental Protection Agency (EPA). (2008a). *Radiological Laboratory Sample Analysis Guide for Incidents of National Significance–Radionuclides in Water.* Revision 0. Office of Air and Radiation, Washington, DC. EPA 402-R-07-007, January. Available at:

www.epa.gov/narel/recent_info.html.
[3] U.S. Environmental Protection Agency (EPA). (2008b). *Standardized Analytical Methods for Environmental Restoration Following Homeland Security Events, Revision 4.0*. Office of Research and Development, Washington, DC. EPA/600/R-04/126D, September. Available at: www.epa.gov/ordnhsrc/sam.html.
[4] U.S. Environmental Protection Agency (EPA). (2009a). *Radiological Laboratory Sample Analysis Guide for Incidents of National Significance–Radionuclides in Air*. Revision 0. Office of Air and Radiation, Washington, DC. EPA 402-R-09-007, June. Available at: www.epa.gov/narel/recent_info.html.
[5] U.S. Environmental Protection Agency (EPA). (2009b). *Method Validation Guide for Radiological Laboratories Participating in Incident Response Activities*. Revision 0. Office of Air and Radiation, Washington, DC. EPA 402-R-09-006, June. Available at: www.epa.gov/narel/recent_info.html.
[6] U.S. Environmental Protection Agency (EPA). (In preparation). Guide for Radiochemical Laboratories for the Identification, Preparation, and Implementation of Core Operations Unique to Radiological Incident Response. Revision 0. Office of Air and Radiation, Washington, DC.
[7] Mann, W.B., A. Rytz, and A. Spernol (1991). *Radioactivity Measurements: Principles and Practices*. Pergamon Press, p. 65
[8] *Multi-Agency Radiological Laboratory Analytical Protocols Manual* (MARLAP). (2004). EPA 402-B-04-001A, July. Volume I, Chapters 6, 7, 20, Glossary; Volume II and Volume III, Appendix G. Available at: www.epa.gov/radiation/marlap.

End Notes

[1] Throughout this guide, the term "Incident Commander" (or "IC") includes his or her designee.
[2] The sum of the effective dose equivalent (for external exposure) and the committed effective dose equivalent (for internal exposure). TEDE is expressed in units of sievert (Sv) or rem.
[3] The common unit for the effective or "equivalent" dose of radiation received by a living organism, equal to the actual dose (in rads) multiplied by a factor representing the danger of the radiation. "Rem" stands for "roentgen equivalent man," meaning that it measures the biological effects of ionizing radiation in humans. One rem is equal to 0.01 Sv.
[4] The term "analytical action level" (AAL) is used in this publication series as a general term denoting the radionuclide concentration at which action must be taken by incident responders. The AAL should always correspond to a PAG or risk-based dose.
[5] See Appendix IV for further references to how measurements are used to make decisions regarding PAGs and action levels.

[6] In this context, sensitivity refers to the ability of the screening equipment to detect different particles.
[7] Knoll, Glen F. 1979. *Radiation Detection and Measurement*, New York: John Wiley and Sons, Inc.
[8] ASTM International (ASTM D7282-06). *Standard Practice for Set-up, Calibration, and Quality Control of Instruments Used for Radioactivity Measurements*, ANNEX X2. West Conshohocken, PA. Available for purchase from: www.astm.org/Standards/D7282.htm.
[9] The volume of 68 m^3 is used as a reference volume as described in the Radiological Laboratory Sample Analysis Guide for Incidents of National Significance–Radionuclides in Air (2009, In Preparation).
[10] See example in Appendix III for ^{192}Ir.
[11] The response curves and characteristics for these instruments were taken from information provided by Ludlum Measurements, Inc., at www.ludlums.com.
[12] FOM = [(detector efficiency)^2/background] (Mann et al., 1991).
[13] A technique that has been used successfully to determine the energy of beta-only emitters is to measure the range of the beta particles in a pure material ("Feather analysis"). The ranges of beta particles in several pure materials (such as aluminum) have already been established. The units of thickness are expressed as areal density, or mg/cm^2. A set of aluminum absorbers of varying thickness is used, and the activity versus the absorber thickness is plotted on a semi-log scale. The linear portion of this curve is then extrapolated to find the "zero" activity thickness. This is then related to the $E_{\beta max}$ of the beta particle, which will be characteristic for a particular radionuclide. A discussion of this technique can be found in Chase, G.D. and J.L. Rabinowitz (1967). *Principles of Radioisotope Methodology*, 3rd Edition. Minneapolis: Chase and Burgess.
[14] The date of this change and the reason for the change need to be noted in the instrument data files.
[15] It is important to note that the use of this type of curve is not necessary for alpha instruments since the alpha response would be mostly independent of energy.
[16] Although the ^{99m}Tc ($t1/2$ = 6 hours) and its ^{99}Mo ($t1/2$ = 66 hours) precursor have decayed, the progeny ^{99}Tc has a half-life of 2. 1× 10^5 y, and will thus be present in the environmental samples exposed during the accident.
[17] Note that the energy of ^{241}Am (59 keV) is somewhat lower than that of ^{57}Co (122 and 135 keV) and will be significantly affected by the aluminum shielding on the NaI(Tl) detector.
[18] The laboratory homogenized the samples by shaking prior to opening and performing the gross screen. The values will be affected due to sample self-shielding.
[19] The gross gamma estimate for the entire sample is 19,000 pCi. This gives about 19,000/250g = 76 pCi/g. Taking a 2-g sample would result in 152 pCi, exceeding the laboratory limit. An aliquant of 1.3 g yields 98 pCi.
[20] Calibration points for the curve were 88, 320, 662, 1115, and 1836 keV. The standards were counted for 5 minutes each in a shielded geometry. The standards used were individual radionuclides (i.e., *not* a mixed gamma ray source).
[21] Ir-192 has several different gamma rays. The average energy per decay event is approximately 390 keV based on the sum of the gamma ray abundances multiplied by their respective energies.

INDEX

A

activity level, 12, 14, 26, 59, 60, 61, 62, 63, 65, 78, 88, 89, 90, 93, 111
adjustment, 16, 97, 114
alpha activity, 97, 98, 109
alternative hypothesis, 72, 74
analytical action level, 2, 9, 12, 68, 69, 70, 71, 72, 75, 80, 84, 122
analytical decision level, 69, 70, 80, 90
analytical protocol specification, 2, 6, 69, 74, 76
aquifer, viii, 83
assessment, vii, 85, 93, 109
atmosphere, viii, 83
authorities, 68

B

background radiation, 69
beta particles, 85, 123
bias, 3, 6, 11, 12, 17, 18, 22, 23, 26, 35, 36, 37, 38, 39, 40, 53, 54, 55, 56, 57, 58, 59, 62, 63, 64, 65, 68, 69, 75, 78, 98
bounds, 53, 58

C

calibration, 70, 85, 86, 87, 88, 92, 98, 99, 101, 104, 109, 113, 114, 115, 116, 119, 120
cation, 16
cellulose, 21
CERCLA, 80
cesium, 100
chemical(s), 2, 6, 7, 8, 15, 17, 18, 28, 33, 34, 68, 70, 72, 77, 80
cleanup, 84
Code of Federal Regulations, 80
coefficient of variation, 77
combined standard uncertainty, 2, 12, 49, 50, 51, 70
compaction, 104
composition, 11, 96
Comprehensive Environmental Response, Compensation, and Liability Act of 1980, 80
Compton effect, 95
configuration, 100, 101, 104, 105, 107, 108, 110
constituents, 6, 17
construction, 21, 69
consumption, 42, 43
containers, 86, 101
contaminant, 81, 113
contamination, 20, 84, 85, 92, 114

control condition, 75
correction factors, 87, 97
critical value, 38, 60, 61, 62, 64, 70, 77
crop, 118

D

danger, 75, 122
data collection, 71
data quality objective, 2, 71, 73, 80
data transfer, viii, 84
database, 111
decay, 2, 15, 19, 42, 43, 44, 45, 46, 47, 75, 80, 89, 90, 93, 97, 101, 109, 111, 117, 123
decontamination, 18, 21, 22, 77
Department of Energy, 80
derived air concentration, 2, 29, 71, 80, 88
derived radionuclide concentration, 2, 13, 14, 18, 20, 26, 27, 48, 69, 71, 84, 88, 89
derived water concentration, 2, 68, 71, 75
detectable, 3, 6, 8, 10, 11, 56, 71, 74, 76, 81
detection, 6, 10, 11, 14, 17, 56, 70, 71, 74, 82, 86, 89, 90, 91, 94, 95, 99, 100, 101, 104, 105, 106, 109, 114, 116
detection system, 94
detection techniques, 109
detonation, 90
deviation, 11, 57, 58, 59, 63, 65, 69
digestion, 19
discharges, 21
discrimination, 72, 80, 88, 93, 96, 100, 113, 114
dispersion, vii, viii, 5, 10, 77, 83, 89
distribution, 33, 35, 37, 38, 60, 61, 112
drinking water, 113

E

effluents, 21
electron, 3, 80, 81, 90
emergency response, vii, 83, 92, 100, 114
emission, 15, 19, 29, 80, 90, 91, 92, 93, 97, 109, 111
emitters, 99, 101, 109, 113, 116, 123
energy, 15, 72, 80, 81, 85, 86, 88, 92, 93, 94, 95, 96, 99, 101, 103, 104, 109, 111, 116, 120, 123
environment, vii, 69, 83
environmental conditions, 16
environmental contamination, vii, 4
Environmental Protection Agency(EPA), v, vii, viii, 1, 2, 4, 5, 6, 8, 16, 29, 40, 41, 48, 49, 77, 79, 80, 83, 84, 90, 110, 113, 121, 122
equilibrium, 101, 117
equipment, 85, 86, 87, 89, 92, 99, 104, 105, 106, 110, 111, 113, 114, 115, 118, 123
evacuation, 84
evidence, 74, 120
experimental design, 16
exposure, 71, 75, 77, 84, 88, 112, 122
extraction, 15

F

false negative, 77
false positive, 77
federal agency, 82
FEM, 40
fiber, 21
figure of merit, 80, 104
filters, 12, 13, 18, 21, 27, 38, 95, 96, 97, 113, 115, 119, 120
flexibility, 69
food, 84
formula, 25
freedom, 34, 37, 38, 60, 61, 62, 77

G

gamma radiation, 72, 76, 94, 101, 106
gamma rays, 94, 117, 123
gamma spectrometry, 81
gel, 21
geometry, 87, 88, 93, 99, 101, 103, 104, 105, 106, 108, 111, 113, 115, 117, 119, 120, 123

Index

germanium, 80, 81, 82
GPC, 81, 86, 97, 100, 109, 113
guidance, vii, viii, 5, 6, 14, 17, 34, 83, 85, 86, 87, 113

H

half-life, 15, 82, 123
halogen, 101, 103
Holst-Thyregod (H-T) test, 62, 65
housing, 69, 95
human, vii, 4, 72, 84
human exposure, 84
human health, vii, 4
hypothesis, 35, 52, 72, 74, 77, 81
hypothesis test, 72, 74, 77

I

identification, viii, 15, 40, 83, 84, 85, 90, 95
identity, 15, 19, 86, 88, 111, 114
improvised nuclear device (IND), viii, 81, 83, 90
Incident Commander (IC), viii, 2, 10, 12, 18, 22, 26, 30, 35, 36, 48, 51, 77, 81, 83, 85, 87, 88, 89, 113, 118, 119, 120, 122
incident of national significance (INS), vii, 81, 83, 86, 87
individuals, 1, 75, 79
initiation, 33
integrity, 85, 86
intentional release of radioactive materials, viii, 83
International Organization for Standardization (ISO), 3, 8, 10, 11, 40, 41, 73
ionization, 99
ionizing radiation, 72, 76, 122
isolation, 8
isotope, 70
issues, viii, 20, 83, 87, 92

L

lead, 1, 79, 109, 116
life cycle, 73
linear function, 12
liquids, 12, 14, 21

M

magnitude, 11, 13, 23, 36, 55, 57, 109
MARLAP method, 22, 24, 57
mass, 14, 70, 85, 86, 104, 109
materials, vii, viii, 2, 3, 4, 11, 13, 15, 20, 23, 69, 74, 81, 83, 88, 95, 104, 109, 118, 120, 123
matrix, 5, 6, 12, 13, 14, 15, 17, 18, 19, 20, 21, 22, 23, 24, 26, 27, 28, 30, 32, 33, 34, 35, 39, 48, 50, 53, 68, 69, 71, 73, 74, 75, 88, 90, 95, 106
matter, iv, 72, 109
maximum contaminant level, 81, 113
measurement, 3, 5, 6, 7, 8, 9, 10, 11, 15, 25, 35, 37, 38, 39, 50, 57, 58, 60, 61, 62, 63, 68, 69, 70, 71, 72, 73, 74, 75, 76, 77, 81, 85, 88, 89, 93, 95, 98, 99, 104, 105, 106, 107, 113, 114, 116
measurement bias, 62
measurements, 11, 25, 28, 56, 57, 59, 62, 63, 75, 76, 87, 88, 91, 92, 93, 95, 101, 104, 109, 111, 112, 121, 122
media, 14, 18, 21, 28, 29, 46, 47, 68, 84, 85, 87
metals, 21
meter, 3, 93, 99, 100, 103, 111, 112, 119
methodology, 87
minimum detectable concentration, 3, 6, 8, 10, 74, 76, 81
Multi-Agency Radiation Survey and Site Investigation Manual, 3, 41, 81
Multi-Agency Radiological Laboratory Analytical Protocols Manual, 3, 6, 41, 56, 81, 122
multiplier, 25, 60, 61, 62, 64

Index

N

normal distribution, 25, 60
nuclides, 27, 46, 47, 91, 96
null, 34, 72, 74, 77, 81
null hypothesis, 34, 72, 74, 77, 81

O

operations, 70
organic compounds, 7
organism, 75, 76, 122
oxidation, 15

P

parallel, 13, 105
peer review, 2, 79
percentile, 42, 43
pH, 16
photons, 93, 94, 109
physical properties, 94, 103
population, 11, 72, 75, 76, 84, 89
positron, 90
precipitation, 8
preparation, iv, 6, 7, 8, 29, 69, 73, 76, 77, 122
probability, 2, 11, 22, 23, 33, 34, 36, 53, 54, 55, 60, 65, 71, 74, 77, 91, 95
probe, 100, 103
proficiency testing, 17
project, vii, 1, 5, 6, 7, 10, 12, 13, 14, 15, 16, 17, 22, 26, 27, 28, 30, 32, 33, 34, 36, 39, 63, 65, 68, 69, 72, 73, 74, 75, 76, 79, 98, 110
protection, 75, 84, 85
protective action guide, 3, 9, 69, 71, 75
purity, 81, 100

Q

quality assurance, 82
quality control, 3, 12, 70, 75, 82
quantification, viii, 15, 24, 83, 85, 90, 91

R

radiation, 2, 3, 8, 40, 41, 69, 70, 72, 75, 76, 79, 81, 82, 84, 85, 86, 88, 89, 92, 93, 94, 95, 102, 104, 109, 121, 122, 123
radio, viii, 83
radioactive contamination, 76, 88
radioactive waste, 118
radiological dispersion device (RDD), vii, viii, 3, 5, 82, 83, 90
Radiological Laboratory Sample Analysis Guide for Incidents of National Significance, 10, 13, 18, 19, 24, 26, 27, 29, 40, 48, 49, 77, 89, 90, 121, 122, 123
radon, 7
reading, 114, 119
real time, 92
reasoning, 97
recommendations, iv, 86, 92, 114
recovery, vii, viii, 5, 18, 83, 84, 118
rejection, 24, 60, 62, 77
relative standard deviation (RSD), 3, 65
remediation, 84
requirements, viii, 5, 6, 10, 17, 22, 27, 28, 30, 32, 33, 69, 73, 74, 75, 77, 83, 84
resources, viii, 83, 84, 85, 100
response, vii, 5, 6, 7, 9, 10, 11, 12, 13, 14, 16, 17, 18, 20, 26, 30, 33, 34, 36, 39, 72, 76, 83, 85, 86, 87, 88, 89, 90, 93, 94, 95, 96, 97, 101, 103, 104, 109, 110, 111, 115, 116, 120, 123
risk, 9, 13, 14, 18, 19, 26, 29, 46, 47, 48, 69, 71, 84, 88, 89, 92, 98, 113, 122
roentgen, 3, 4, 72, 76, 82, 83, 122
root, 22, 56, 57, 58, 59, 60, 62
runoff, 48, 50, 51

S

safety, 20, 87
sample survey, 93
scattering, 108

scope, 7, 14
sediment, 21, 27
sediments, 13, 31
selectivity, 14
semiconductor, 80
sensitivity, 92, 106, 123
significance level, 37, 38, 60
sodium, 81, 109, 116
solubility, 44, 45, 46, 47
solution, 16, 70
species, 15, 17, 34
specifications, 5, 14, 20, 24, 26, 33, 74, 76
stability, 16, 73
staff members, 110
standard deviation, 3, 9, 25, 33, 37, 38, 39, 51, 56, 57, 58, 59, 62, 65, 75, 77, 112
standard error, 77
statistics, 56, 62, 64
strontium, 16
style, 106, 108, 120
Superfund, 80
supervisor, 119
suppliers, 120
surface area, 13
surveillance, 17
suspensions, 21

T

target, 9, 15, 60, 68, 69, 73, 74, 76
technical definitions, 68
techniques, viii, 75, 83, 85, 86, 98, 113, 115
terrestrial areas, viii, 83
test statistic, 60
testing, 3, 17, 23, 26, 27, 28, 30, 32, 33, 34, 35, 37, 38, 48, 57, 74

thallium, 81
thorium, 101
time frame, 84
tissue, 72
training, 33, 110
transportation, 114
treatment, 21, 22, 68
turnaround time, viii, 6, 82, 83, 92
Type I error, 33, 34

U

uniform, 112
upper bound of the gray region, 82

V

validation, vii, 3, 5, 6, 7, 8, 10, 11, 12, 14, 15, 16, 17, 18, 19, 20, 21, 22, 23, 24, 25, 26, 27, 28, 30, 32, 33, 34, 35, 36, 37, 38, 39, 48, 53, 54, 55, 56, 57, 58, 59, 62, 63, 65, 68, 71, 72, 74, 75, 77, 87, 110
vapor, 46, 47
variables, 16
variations, 16, 21, 73, 113
vegetation, 21

W

water, viii, 2, 5, 10, 13, 18, 19, 20, 21, 22, 26, 27, 28, 29, 31, 37, 38, 42, 43, 48, 50, 51, 53, 68, 71, 75, 83, 84, 113
wood, 21